GOLD PLACERS & MINERAL DEPOSITS
THEIR FORMATION, DEPOSITION, AND CHARACTERISTICS

VOLUME 1

PINE NUT PRESS
Minden, Nevada

Copyright © 2018 Susan Lee Parkhurst

All rights reserved.

ISBN: 1978449623
ISBN-13: 978-1978449626

Library of Congress Control Number:
2017960233

The Dave W. Parkhurst Mining Writing Collection
was compiled, edited and designed
by Susan Lee Parkhurst
and produced by Pine Nut Press.

Note to Reader

The content of this book is the work of late mining writer/consultant Dave W. Parkhurst (deceased, 1993). The articles contained herein are reprinted from the *California Mining Journal* (now the *International California Mining Journal,* or *ICMJ*) with permission from the *ICMJ* publisher. No substantive changes have been made to the text in the process of digitizing the material for publication in this book, and no liability is assumed for any inaccuracies that may have occurred in the process of converting the printed magazine articles to the digital format, or from the digital format to the printed book. Further, the content reflects the state of the mining industry and its technologies and processes, as well as the knowledge base that existed, as of the time the articles were published in the *CMJ*. No guarantee is made as to the accuracy or current relevance of any of the content.

To
David Walter Parkhurst
in loving memory
and
appreciation

Contents

Foreword vii
Editor's Preface ix
Editor's Acknowledgements x
Introduction 3

PART I: HOW AND WHY MINERAL DEPOSITS FORM
1. The Formation of Mineral Deposits 7
2. Metallic Differentiation in Magmas 11
3. Alluvial Fan Gravels 15
4. The Formation of New Placer Gold Deposits 19
5. New Gold Crops after Heavy Runoff 25

PART II: PLACER DEPOSIT FORMATION AND CHARACTERISTICS
6. Placer Characteristics in the Great Basin and California 33
7. Alluvial Placers 39
8. Beach and Marine Placers 45
9. Desert (Dry) Placers and Gold Deposition 49
10. Glacial, Eolian, and Ancient Placer Deposits 55
11. Residual and Eluvial Placers 59
12. Buried Placer Deposits 65

PART III: GOLD, SILVER, PLATINUM, AND DIAMONDS
13. The Geology and Technology of Gold 71
14. Gold: Its Character and Concentration 75
15. Types of Gold Ore Deposits 79
16. Separating and Identifying Gold Particles 83
17. Placer Gold Mining 89
18. The Silver Minerals 95
19. The Mineralogy and Geology of Platinum Deposits 99
20. Prospecting for Diamonds 107

PART IV: OTHER METALS AND MINERALS WORTH SEEKING
21. Beryllium: The Space Age Metal 113
22. Chromium and Chromite 115
23. Cobalt: A Strategic Metal 119
24. Mercury and Cinnabar 123
25. Tantalum and Columbium 129
26. Tellurium: Properties, Ores, and Assay Procedures 135
27. The Geology and Technology of Thorium 141
28. The Titanium Minerals 145

29	Tungsten Deposits	151
30	Zirconium and Hafnium	155
31	The Rare Earth Metals and Minerals	159

Afterword 165
About the Articles' Author..167
Dave W. Parkhurst Mining Writing Collection (Appendix A) 171
Year of Publication in the *CMJ* (Appendix B)......................................173
Publications Referenced *(selected references)* (Appendix C)................175
References *(works cited)* 177
Index..179
Editor Contact Information 188

Foreword

In 1987 I was appointed by then-Governor Richard Bryan to the post of Executive Director of the Nevada Department of Minerals. The Department (now Division of Minerals) and the Nevada Commission on Mineral Resources is a state agency whose mission is to "promote the responsible development of mineral resources in Nevada." Shortly after I started with the State, I was attending a Nevada Miners and Prospectors Association lunch meeting at Grandma Hattie's restaurant in Carson City. It was there that I had the good fortune to meet a man who, I quickly learned, was one of the most rational and effective voices for the small-scale miners and prospectors in Nevada and the west, Dave Parkhurst. Dave became my friend and we worked together as advocates for responsible mineral development many times on many issues until his untimely passing in 1993. Dave's work and steadfast reporting through his articles compiled by Sue Parkhurst in *Fighting the Good Fight* (Volume 4 of the four-volume Dave W. Parkhurst Mining Writing Collection) reflect much of that work.

The late 1980s and early '90s was a time of rapid growth in Nevada's mining industry with the discovery, development and expansion of major gold mines in the northern and central part of the state. Large multinational mining companies were doing much of this development, but even with these well-funded industry players, the small-scale miner and prospector played the role that they always have throughout time. As ever, in many cases, original discoveries of mineral deposits were made and mining rights obtained by much smaller entities, and then later sold to the larger companies.

About 86 percent Nevada's surface was then and is now largely managed by the federal government through its US Bureau of Land Management and US Forest Service. This fact, together with the growth of major mining activities in Nevada and the west in general, more than ever attracted the attention of legislators, regulators and the public. Unfortunately, many individuals from these groups would prefer that mining with its impacts on the environment not take place. This is despite the fact that all people, regardless of where they live or how they live need the products made from a variety of minerals that have to be mined from the earth. When you think about it, you'll find the phrase "If it isn't grown, it has to be mined" is true.

By 1990, the stage was set for many changes to the federal and state laws, regulations and policies that governed mining and mineral exploration activities on public land, later moving to private land. Unless there were strong mining advocates with clear and concise positions effectively brought to the state legislatures and the US Congress, mining for hard rock minerals in the west could have been irreparably damaged. The large mining companies had the ability and wherewithal to affect this process to reasonably manage the impact on their businesses, but without people like Dave Parkhurst, it was easy to see the small-scale miner and prospector as an endangered species. While still very challenging, that didn't happen. People still prospect, stake mining claims and make important contributions to keeping the supply of mined products flowing.

Today, while always a challenge in terms of public debate, mining on both public and private land is proceeding in a safe and environmentally responsible manner while having significant economic impacts. This is thanks, in part, to individuals like Dave Parkhurst.

RUSS FIELDS
Reno, Nevada

Russ Fields worked in the mining industry and was executive director of the Nevada Department of Minerals, president of the Nevada Mining Association, and, most recently, director of the Mackay School of Earth Sciences and Engineering (formerly Mackay School of Mines), University of Nevada, Reno. He is now retired.

Editor's Preface

The four volumes that comprise the Dave W. Parkhurst Mining Writing Collection, of which this volume is the first, are the culmination of a project to preserve the written works and legacy of mining writer, consultant, advocate, prospector and exploration professional David Walter (Dave W.) Parkhurst. This compilation of Dave's writing is intended by me, his widow and the publisher/editor of the collection, as a tribute to him and the fulfillment of a promise I made to myself in the years following his death. That promise was to honor Dave by ensuring that an important part of his legacy lived on through preservation of his work in book form.

The two of us had discussed the possibility of Dave's writing a book someday, but time ran out before he could achieve that goal when a heart attack claimed his life in 1993. Although the "DWP" Mining Writing Collection is no substitute for the actual book, or books, Dave might have written, it does provide a compilation of much of his best work and makes the knowledge, expertise and insight that he acquired over the years available to posterity.

Dave wrote on multiple mining-related topics—from the formation, concentration and characteristics of gold deposits, to how and where to prospect for the precious yellow metal, to new technologies in the mining industry, to political and governmental actions affecting miners and prospectors, to the negative impacts of anti-mining propaganda on the public's perception of mining and miners, to other important topics concerning the mining industry—particularly small-scale mining and prospecting in the American West.

Many of Dave Parkhurst's readers appreciated his ability to convey technical information clearly and comprehensibly. Among these readers were many who regarded him as a trusted expert in various facets of mining and prospecting, but he didn't take the "expert" moniker too seriously. "The more I know, the more I know I don't know," he would say with characteristic humility and wry humor.

Writing about mineral exploration, prospecting and mining was not nearly as enjoyable for him as being in the field actually doing those things, but he was gratified that many individuals appreciated and benefited from the knowledge he imparted over the years and from his fight to preserve the rights of miners in this country.

Editor's Acknowledgements

Many thanks to Scott Harn, publisher of the *International California Mining Journal (ICMJ*, formerly the *CMJ*) for his permission to reprint Dave Parkhurst's *CMJ* articles in the four volumes that make up the Dave W. Parkhurst Mining Writing Collection.

I could not have completed this work without the fantastic efforts and cooperation of my siblings and their families in caring for our aging parents so beautifully in the final years and days of their lives. Special thanks to Sheri, Lisa, Lori, Robin, Bill, and Bob. Posthumous thanks to my parents, Evelynne and Leonard, for everything.

I am also deeply appreciative of the excellent care given to my mother, an Alzheimer's patient, by Debi Stryker and her staff at Aunt Dottie's Place over the past several years. Knowing she was in such good hands made it possible for me to concentrate on my work and devote to this project the many hours required to see it through.

My sons and Dave's, Dan and Nick, contributed in important ways that allowed me to focus my efforts on bringing forth this compilation. Dan, an IT professional, kept my computers and other office equipment running smoothly (as his dad used to do back in the day). My thanks and appreciation to both of them and also to my little grandson, Ethan, for giving up some of Nana's time and attention while she worked on this project in remembrance of the grandfather he never knew. I hope this work will someday help him know more about the person "Grandpa Dave" was.

Judy Haar, my former SCORE-Reno mentor, was gracious with her time and professional expertise when I was trying to define what it was I wanted to produce as an author and publisher, and throughout the process of writing and publishing my first book.

Thanks also to Dana Bennett, president of the Nevada Mining Association, for taking time to meet with me and offer valuable suggestions, feedback, and encouragement regarding my "Dave books" project. I appreciate, as well, NvMA Office Manager Lauren Arends' efforts in facilitating my contacts with Dana.

Finally, my thanks to Russ Fields for providing the foreword to the Dave W. Parkhurst Mining Writing Collection. Dave would have been pleased and honored by Russ's contribution to this collection.

GOLD PLACERS & MINERAL DEPOSITS
THEIR FORMATION, DEPOSITION, AND CHARACTERISTICS

VOLUME 1

Dave W. Parkhurst
MINING WRITING COLLECTION

Introduction

As mentioned previously, the Dave W. Parkhurst Mining Writing Collection consists of articles originally published in the *California Mining Journal*. This book, the first volume in the collection, consists of material that relates to the formation, concentration, characteristics and types of deposits of valuable metals and minerals—especially gold, historically the most sought-after of the precious metals. The other three volumes are described in the back pages of this book.

It should be kept in mind when reading the articles in this compilation that they were written over a period of years spanning 1982 to 1993, and significant changes have occurred in several aspects of the mining industry since then that can make some of the content obsolete in some respects. This is particularly true about the many advances in mining exploration technology that have been made since Dave's passing but also to changes in the political and economic landscapes, to ongoing fluctuations in the precious metals prices and market conditions, and to the impact of the Internet on the accumulation and distribution of information.

That being said, a good deal of the content in this collection of Dave's mining writing, most notably in books one and two, is timeless (for example, geological processes and their effects on mineral formation) or otherwise still relevant today (as is the issue of how much wilderness is enough, for instance). Besides having experience and expertise in the field, specializing in mineral exploration, Dave was well-versed in geology, mineralogy, chemistry, engineering, technology and other scientific and technical disciplines—as well as the status of wilderness withdrawals, the wetlands issue, excessive governmental regulation, and other issues concerning the natural resource industries and public lands use. He wrote extensively on these subjects and others, in particular as they relate to mining and prospecting for metals and minerals.

Dave, who wrote prodigiously under often quite challenging circumstances, was highly accurate in the copy he cranked out on his little Smith Corona typewriter. His typed drafts were usually typeset for print practically as-is—that is, with a minimum of editing and proofreading required. In the process of digitizing and preparing the articles from Dave's archive of old CMJs, a few typos, grammatical errors and style inconsistencies were detected and corrected; no substantive or significant changes were made to his writing.

My apologies as editor and publisher of this collection for any errors that may have crept into any of the articles in the process of digitizing them and otherwise preparing them for printing. Great effort has been made to preserve the integrity of Dave's written work.

A final note about the organization, content, and editing of this compilation: The four volumes that make up the collection consist of about 150 journal articles, in total, written over an 11-year period on a variety of mining-related topics. They have been organized as logically as possible (given the constraints of time) — sometimes in chronological sequence, other times by some other "content relatedness" criteria. A number of the articles contain some content that is duplicative of, or closely resembles, some of the content in other articles written on the same or a similar topic; such articles generally included updated information or material that was supplemental or perhaps written from a different slant. Also, as previously noted, some of the content has become obsolete since it was originally published. It was beyond the scope of this project to eliminate entirely such duplicative or obsolete material.

PART I

How and Why Mineral Deposits Form

❝ [T]he transportation and deposition of minerals is a very complex process that never really stops. New minerals are continually being formed as older minerals are decomposed, and the decomposed minerals are constantly being recycled over and over again by the changes in geological conditions operating over extremely long periods of time. ❞

1
The Formation of Mineral Deposits

THE PHYSICAL AND CHEMICAL PROPERTIES of elements largely determine whether or not they occur together or in association with each other in the same geological environment. Those elements having the same or similar physical and/or chemical properties commonly occur together in mineral deposits; those having a dissimilar nature generally do not.

Minerals and ores are formed under a vast number of varying geological conditions and are then often subsequently altered, removed and redeposited, enriched, or dissolved and partially or wholly dispersed. However, there are certain mineral combinations that are characteristic of a particular set of geological conditions. This means that specific groups of related elements often occur together in certain types of geological formations, whereas they are seldom found together in other types of formations except under highly unusual circumstances.

The geochemical and geological conditions to which the elements are subjected largely determine both the method and extent of their deposition. The geological forces that control the existence of each particular mineral environment include such factors as temperature, pressure, the relative abundance and properties of the elements present, and the presence or absence of various solvents and/or catalysts.

In most cases, the original mineral deposits will have undergone extensive alteration due to subsequent changes in the geological environment over time. These changes result from the constantly shifting forces in the earth's crust. The following general description of the various processes involved in mineral transportation and deposition may assist in the location and identification of mineral deposits in the field.

Those elements having a low solubility and volatility usually tend to remain fairly close to their original plutonic source. Since all magmas contain accompanying gases, these elements commonly travel short distances into the surrounding rocks, resulting in the formation of mineral veins and pockets. This group of elements combines into high-temperature minerals that form deep-seated veins similar to, and in association with, the granitic pegmatites.

In this stage, all of the uncombined elements are still concentrated in a residual liquid phase. This liquid is still very fluid, so minerals such as quartz, feldspar and mica tend to crystallize in coarse masses that contain associated elements, such as beryllium, boron, tin, tungsten, fluorine, tantalum, columbium and several similar elements. Pegmatites formed in this environment sometimes also contain commercial deposits of mica, beryl and the rare earth metals.

After the magma has cooled, become less fluid, and deposited the high-temperature minerals, there are still considerable amounts of re-condensed, hot watery solutions present in the mass. This solution contains dissolved metals, silicon, sulfur and other elements. As this liquid escapes into areas of lower temperature and pressure, the dissolved elements combine and form mineral deposits along the fissures through which the solution passes. The lower-temperature veins that are formed in this manner frequently contain valuable ore minerals, including most precious metal vein deposits.

The rocks bordering these fissures are often altered by the escaping mineral solutions, with some of the ore minerals replacing more soluble substances in the wall rocks. Under certain conditions, a higher concentration of ore minerals is deposited in the wall rocks than is present in the mineral vein that originally contained them.

In those areas where recent volcanic activity has taken place, the water-soluble elements commonly travel in solution for great distances from their original source, eventually being deposited at much lower temperatures and pressures closer to the ground surface. Minerals such as stibnite (antimony sulfide), cinnabar (mercury sulfide), argentite (silver sulfide), and metallic gold are deposited under these conditions.

The resulting veins and their mineral-enriched borders are also affected by exposure to air and moisture near the ground surface. Most of the sulfides will be oxidized to form water-soluble sulfates. Some of these sulfates are removed in solution while a portion will sink downward and react with other sulfides, thereby enriching certain zones by replacing some of the other elements. Some of the sulfides will also be altered to carbonates, silicates or oxides by reactions with the wall rocks. A new group of ore minerals can be formed and sometimes enriched by this process, or the near-surface outcrops may be leached of their mineral content.

The Formation of Mineral Deposits

Disseminated metallic elements may also become concentrated as hot mineral solutions seep through permeable sedimentary rocks, eventually forming massive ore bodies. The metals sometimes combine with sulfur and separate into cracks and crevices to form ore veins. The low-temperature sulfates, sulfides and carbonates are quite commonly found in fissures and cavities in sedimentary rocks. Likewise, veins filled with low-temperature sulfides, quartz and calcite are also found in sedimentary rocks.

As rock formations disintegrate and erode away, the more chemically resistant minerals remain behind. Heavier and more durable elements, such as gold, are often left near their original source as the lighter and less resistant minerals are broken down and washed away. The heavier and more durable elements can eventually form economical residual placer deposits by the concentration of mineral particles contained in tremendous volumes of rocks that have eroded away over a period of millions of years. Most of the less chemically resistant minerals are dissolved and removed in solution, and the lighter materials are blown or washed away. Most of the disintegrated materials are eroded downslope to more level terrain where they are deposited in sedimentary formations.

As the sedimentary formations are subjected to changing geological conditions over long periods of time, they are buried, compressed, reheated and sometimes melted. In this process, the secondary minerals are again converted to high-temperature minerals. Variations in this process also create an environment that is favorable to the formation of a new group of minerals called metamorphic rocks.

If molten magmas penetrate into the compressed mass of metamorphosed rocks, additional changes in the composition of the rocks can be produced. A new series of contact-metamorphic minerals can develop, creating yet another mineral-forming environment. These new minerals, in turn, decay and become members of the oxides, thereby forming an entire set of hydrated minerals.

It should now be evident that the transportation and deposition of minerals is a very complex process that never really stops. New minerals are continually being formed as older minerals are decomposed, and the decomposed minerals are constantly being recycled over and over again by the changes in geological conditions operating over extremely long periods of time.

However, a basic familiarity with the various types and methods of mineral deposition can be an invaluable aid in the identification of economical mineral deposits, and it can also be utilized to indicate which minerals are most likely to occur in a particular geological formation in which a mineral of known composition has been identified.

Mineral Deposits, a classic text first published in 1933, was written by Waldemar Lindgren. Dave Parkhurst referenced Lindgren's work often in his writing on the subject of mineral deposits. He owned one of the original copies of the book (published in 1933), and it was one of the treasured works in his library. The book has been reproduced by Wentworth Press and is in the public domain in the U.S. The cover shown here is from the Wentworth Press version of the book.

2
Metallic Differentiation in Magmas

SOME OF THE TRACE or accessory minerals contained in molten rocks, or magma, can become concentrated into ore bodies of sufficient size and richness to constitute valuable mineral deposits. Magmatic ore deposits are characterized by their intimate relationship with deep-seated or intermediate igneous rocks. The ore deposits themselves are igneous rocks whose mineral composition gives them an economic value, and they are the sole source of several strategic and critical minerals—such as chromium and the platinum group metals.

Magmatic ore deposits may constitute an entire igneous mass or certain portions of the structure, or they may form as offset ore bodies since they are basically magmatic products that have crystallized from or within the magma itself. The ore bodies and certain associated types of rocks are also called magmatic segregations, magmatic injections, or igneous syngenetic deposits. The deposits are formed by mineral crystallization and/or concentration by differentiation of intrusive igneous masses.

The parent magmas are masses of molten matter that occasionally form in the earth's crust, and they provide the materials from which igneous rocks are formed. By definition, they are high-temperature mixed solutions of silica, silicates, metallic oxides and dissolved volatile elements and compounds, and, as such, they usually are governed by the same basic laws as ordinary chemical solutions. The composition of magmas is not the same as the composition of the rocks they form, because of the mineral segregations and escape of volatile gasses that occur during the various processes of magma movement, lowering temperature and pressure, and solidification. As a result, several different types of rock and rock formations are usually formed from a single parent magma.

The composition of magmas varies as widely as the large variety of rocks that they form. The dissolved volatiles normally consist of water, boron, carbon dioxide, chlorine, fluorine, sulfur and minor amounts of other substances. The volatiles play an important part in lowering the melting point of the magma, in decreasing its viscosity, in collecting and transporting the contained metals, and in the eventual formation of the resultant mineral deposits. Most of the volatiles escape the solution as the magma solidifies, and the remainder form chemical compounds with metallic ions.

Molten rocks are only temporarily created in localized areas of the planet's crust, and they are usually confined in magma reservoirs, layers or pockets as they are forced upward toward the surface and eventually consolidate. The rock-melting is normally localized, and there is no continuous layer of molten material under the solid crust. Although the rock temperature at depth is well above its melting point (because of the heat within the earth), the enormous pressure exerted by the overlying rock formations prevents melting from taking place. If this pressure is reduced by faulting or buckling of the overlying rocks, however, then the underlying rocks will melt and magma will be produced. This is illustrated by the fact that most of the igneous intrusions and volcanic eruptions occur along the zones of greatest weakness in the earth's crust, where crustal disturbances have relieved downward pressure and created conditions favorable to the formation and movement of magma.

As with any liquid under pressure, molten magma tends to move towards the point of least resistance which, in this case, would be predominantly upwards towards the earth's surface. The formation and movement of molten rocks may be caused solely by a reduction in the downward pressure, or it may be facilitated by plate tectonic movement that "squeezes" the molten mass upwards. As the magma is formed, it may also melt and assimilate the overlying rock formations and gradually eat its way upward. If the molten liquid reaches a point of structural weakness, it may vent as a volcanic eruption. If the magma reaches a region of lower temperature and pressure at depth, it may solidify and form an intrusive such as a batholith or igneous stockwork.

It is now generally accepted that most metallic mineral deposits, and many of the nonmetallic deposits, have resulted from primary igneous activity. In other words, the magmas are the original source of the minerals from which economic mineral deposits are formed. Some of these ore deposits were formed during, or soon after, the solidification of the parent magma, while others have been produced by one or more periods of growth by a combination of geological processes.

It has also been noted that there are a large number of intimate associations between certain metals and minerals and specific types of igneous rocks. As a result, it has been determined that the valuable minerals in most ore deposits, whether hypogene or supergene, came either directly or indirectly from parent magmas. This relationship ties the origin and character of most mineral deposits closely to the character and composition of the parent magma and the processes of mineral crystallization in molten solutions.

Since the magmas are molten solutions that obey the laws of aqueous solutions, the crystallization of the contained minerals is dependent upon their relative solubility. In this case, however, the point at which the various minerals will crystallize is not determined by their temperature of fusion, even though no mineral will crystallize above its fusion point. Mineral crystals in the magma will begin to form when the temperature falls below their individual point of saturation in the solution itself. This is also compounded by the fact that a solution of one constituent in combination with another may produce a lower melting point for both that is less than the melting point for either constituent by itself. As a consequence, a magma may remain molten at a temperature that is well below the melting point of all its constituents. This means that crystals will not form until the temperature falls below the melting point of the mixture itself.

After crystallization begins, the most insoluble minerals will crystallize first, followed by other minerals in a definite order as the temperature continues to drop. This produces a segregation of the various metallic minerals in the solution, and it also gives them an opportunity to concentrate in specific locations within the mass. As the crystallization process continues, the remaining aqueous extracts tend to gather the metals that were initially trace constituents dispersed throughout the molten solution. In other words, a further concentration of the contained metals can now occur.

The cooling of the molten liquid normally is more rapid towards the outer extremities of the mass, so crystallization will occur in these areas first. Since volatiles are usually circulating throughout the solution, this provides an opportunity for the extraction and migration of dispersed metallic ions from throughout the mass towards the extremities or the points where the magma comes into contact with cooler, solid rock.

As these processes continue, the originally homogeneous magma splits up into unlike fractions having their own unique composition. This process is called magmatic differentiation, and it may occur in a single stage or in several stages. Before consolidation and solidification, the separate fractions of the parent magma may undergo further differentiation, forming a series of distinct but related rock types. During the latter stages of this process, the aqueous solutions that have been accumulating the residual fractions and becoming increasingly concentrated with metals tend to be tapped off, and provide the metallic constituents of vein structures and other mineral deposits.

During the combined progression of crystallization and differentiation, some of the metallic elements may be collected and concentrated into separate fractions that can consolidate as either a distinct portion of the igneous mass or as a separate injected body, forming magmatic ore deposits. Therefore, differentiation can not only produce different types of rock from the same parent magma, but it can also form magmatic ore bodies of significant economic value.

Even when these processes do not form valuable mineral deposits, they still supply most of the mineral solutions that are utilized in the formation of the majority of metallic mineral deposits, through enrichment by a variety of geological processes. Since these magmatic solutions provide the metallic minerals found in most economical ore deposits, they should be of particular interest to those people who are engaged in mineral exploration activities.

3
Alluvial Fan Gravels

SINCE THE BULK OF THE WORLD'S precious metals production from placer materials has been extracted from alluvial fan, delta and floodplain deposits, a basic description of the form and character of these deposits, their interrelationship and similarities, and their potential for future exploitation will be covered in the following information. Though these types of gravel accumulations vary considerably in size and configuration, they all basically originated as alluvial fan gravels and will often be referred to as such throughout this article.

Alluvial fan gravels are normally deposited at the mouths of rivers, streams and canyons, but they can accumulate at any location where a watercourse reaches an unconstructed and fairly flat area. This can be immediately below the entrance to a canyon, ravine or gorge, at the point where a river or stream intersects the shoreline of a lake or ocean, as well as where canyons widen and flatten out or streams flow into low-lying valleys. Although the term "fan" implies a certain shape, the geological definition refers more to the fanning out or spreading of placer materials as they are moved from a confined to a relatively unconfined space, and the term can apply to a wide variety of placer formations.

The transition from river and stream placer deposits to alluvial fans, deltas and floodplain gravels can sometimes be difficult to describe, as all of these types of placer accumulations often merge with each other and have no distinctive boundaries. Usually, the fan gravels (sometimes called alluvial cones) are deposited at the first major widening and leveling point in a watercourse (which might also be called a delta), most of the delta deposits (often called fan deltas) are described as the accumulations found at the mouth of a river (where it enters a larger river, bay or seashore), and the floodplain deposit refers to the extended portion of fan gravels or the widely dispersed gravels in low-lying valleys (which might also be composed of multiple-channel river and stream gravels). Often, a river delta is in itself both an alluvial fan and a floodplain.

Because water volume, velocity and pressure determine its ability to transport placer materials, any significant reduction in the force produced by these factors also results in a reduction of the size and quantity of gravels being moved. Any sudden widening or leveling in a watercourse produces a sharp drop in the pressure exerted by flowing water and, as a result, large quantities of gravels spread out and accumulate. The heavier and more durable placer minerals being transported by the water will settle and accumulate much more rapidly than the lighter sand and gravel, and concentrations of these particles will collect near the upper edge of the fan.

The finer and flatter precious metal particles (plus the major portion of the base-metal minerals) will normally remain intermixed with the bulk of the fan gravels, and they are usually spread out over a considerable distance. If present in sufficient average quantity, these smaller particles can constitute the major economic potential in massive fan, delta and floodplain deposits.

Most of the fan gravel accumulations are fairly permanent, unless subsequent geological uplift or increased water runoff create the conditions necessary for further erosion and movement of the material. Those deposits located on a lake or seashore are also subjected to further alteration by wave action, tidal action, or underwater currents, and may eventually be deposited as beach and marine placers.

Many of the bench and terrace gravel deposits are actually remnants of older alluvial fan gravels, where regional uplift and increased water flows caused the erosion of the major portion of the deposit, leaving some of the gravels at higher elevations.

Because of variations in the rate of uplift or subsidence (or combinations of both) throughout extended periods of time, concentrations of placer minerals in older fan gravel deposits can occur in practically any strata—from near-surface to over a thousand feet in depth—depending upon the conditions existent at the time of deposition. Concentrations can also occur over a considerable lateral width, due to tilting of the terrain, the formation of multiple-channel systems, natural obstructions or wide variations in the rate of erosion. In many instances, the present visible surface indications and topography have very little or no relationship with the conditions that existed at the time the deposit was formed.

As with any other placer concentration, the quantity of valuable placer minerals available for deposition and the length of time in which they are allowed to accumulate will be a determining factor in whether an economic placer can be formed. The formation of the gravel structure itself is dependent upon a number of conditions related to the combined sorting actions of flowing waters and gravity movement, stream gradient, obstructions, regional uplift or subsidence, etc. All of these factors will undergo considerable fluctuations over an extended period of time, so minable placers can be formed over very short periods of time or by the constant accumulation in concentration of values over millions of years.

The above information illustrates the extremely complex nature of placer formation in massive gravel structures over time. There are really no "typical" alluvial fan, delta or floodplain deposits: each deposit, and its contained concentrations of valuable minerals, is somewhat unique, according to the mineralization present in the eroded materials and the topography and climate of the region where (and at the time) it is formed. Even with the wide variations in deposition, however, the general characteristics of most alluvial fan deposits are somewhat similar:

1) Most fan, delta and floodplain deposits are massive accumulations of individual gravel layers (or mud, silt and sand layers in some river deltas), interspersed with a multitude of older, buried stream channels—from the surface to a considerable depth. The old stream channels in the gravel mass would be analogous to the vein system in the human body.

2) The heavier mineral particles and larger gravels are deposited near the upper margin of the gravel deposit, though they might be buried by accumulations of silt, sand and mud. They can also be concentrated in some of the buried stream channels, over fairly long distances under favorable conditions.

3) Finer, lighter, flatter and flakier mineral particles are normally distributed throughout the gravel layers, although some of them may also be concentrated in the buried stream channels.

In wetter climates (at the time of deposition), most fan gravels are well-rounded and smooth, and they are accompanied by large quantities of finely ground sand and silt. In drier climates, most of the gravels are angular and subangular, and are usually mixed with fairly small quantities of sand and silt. As a rule, the heavier placer minerals are also well-mixed in the bulk of the fan gravels in dry climates, due to sudden surges in runoff and flash flooding. As a result, there is very little opportunity for sorting action, accumulation and concentration to take place except in confined channels formed during extended wet periods.

Many of the massive accumulations of alluvial gravels have been found to contain a sufficient quantity of valuable placer minerals to warrant large-scale mining operations. At several times in the past, these deposits have provided the major portion of the world's gold and platinum group production and, at today's prices, they have the potential to again be a major precious metals source. In addition, they might also represent a major source for several other placer minerals, many of which were ignored or discarded during most precious metal placer mining activity.

Though gold and platinum are often the most sought-after placer metals, there are a large number of other placer minerals in demand today. These include such minerals as cassiterite, chromite, columbite, gemstones, ilmenite, monazite, rutile, scheelite, tantalite, wolframite, zircon and others. Sufficient concentrations of these minerals in large volumes of gravel could represent a high potential for profitable mining operations.

The potential for localized concentrations of placer minerals in both small and large fan gravel deposits should also be considered, particularly in buried stream channels. It is also likely that higher relative concentrations of valuable minerals might be found in particular layers (or strata) of alluvial fan deposits—at any depth. Some of these deposits could possibly be mined profitably by open-pit methods (if the stripping ratio isn't too high), or by underground drift-mining methods.

Definitive exploratory drilling or excavation into some of the fan gravel deposits (where valuable minerals are suspected or known to be present) might yield some positive results, especially in highly mineralized districts. With today's modern mining and processing technology, higher average prices, sophisticated exploration techniques, and overall number of individual placer minerals in demand, it seems likely that these placer deposits will represent a significant bulk-minable resource for a variety of minerals in the future.

4
The Formation of New Placer Gold Deposits

ALTHOUGH THEY WERE MINED EXTENSIVELY in the past, many of the larger placer mining districts have a good potential for the formation of newer placer deposits through the natural reconcentration of the remaining gold values and the fairly constant erosion of new gold particles for sorting and deposition. Generally, the longer the time that has elapsed since the area was last mined and the greater the extent of the subsequent erosion, the greater the chances are for the existence of recently formed gold placers. The following information describes the various processes involved in the reconcentration of gold particles into new placer deposits and the characteristics of the resulting placers.

The high potential for the formation of newer placer gold deposits has been referred to by several prominent writers in the mining field, including Waldemar Lindgren in his 1933 text, *Mineral Deposits*. Lindgren stated that many of the major placer mining areas throughout the world would eventually be reworked in the future due to the natural reconcentration and deposition of gold values lost in the original mining operations, the accelerated rate of erosion resulting from the tremendous amounts of gravels that had been disturbed and redistributed by past mining activities, and the ongoing processes of natural erosion supplying more placer gold particles for additional sorting and deposition.

Most of the mining equipment and methods utilized by the early miners were not very efficient and did not recover a high percentage of the total gold values contained in the gravels. This resulted in the loss of significant quantities of gold, especially very fine and flaky particles, in the tailings. High losses were particularly notable in many of the areas where hydraulic mining methods were extensively used, with gold losses in some operations amounting to over 50 percent of the contained values.

The most common type of gold recovery device used in the early days was the sluice box, which will not normally recover much of the smaller, smoother or flattened gold particles efficiently. In addition, many of the sluicing operations performed "cleanup" of the riffles at intervals ranging from a few days to a week or longer. Depending upon the quantity of heavy black sand minerals in the gravels, sluice riffles will tend to "pack" with accumulations of heavy minerals in a fairly short period of time and will not recover any appreciable amount of placer gold while in this condition (except for larger gold nuggets). This practice resulted in a higher-grade concentrate on the average at the expense of efficiency in recovery and consequently produced much higher gold losses in the tailings than those from operations that performed cleanup more frequently.

The prevailing desire to obtain quick riches in the early days of placer gold mining also contributed significantly to higher operational losses, especially among the most inexperienced miners, because almost everyone was looking for the richest gravels and the larger gold nuggets. In many instances, little or no attention was paid to the recovery of the small-to-very-fine gold particles that often constituted the major portion of the total gold values contained in the gravels.

However, many of the major known producing placer districts were reworked by latecomers after the original miners moved on to the newly discovered gold fields. These miners often worked lower-grade deposits left intact by the previous miners as well as some of the tailings left from the earlier mining operations. Some of the miners who reworked these districts, particularly the Chinese, were much more efficient in their mining methods than the original locators.

By this time, however, a tremendous volume of sand and gravel that contained widely dispersed precious metals particles had been spread out and dumped into the rivers and streams. Because most of this material was of marginal or low gold content, only a fairly small portion of the total mass was reworked—even by the patient and thorough Chinese.

There were also large quantities of lower-grade unworked gravels left exposed by the early mining activity and the removal of the old stands of timber for construction materials and firewood. Therefore, the marginal and lower-grade materials that could not be economically mined by the most efficient miners of those days were now stripped of vegetation and exposed to the effects of accelerated erosion. The river and stream beds had also been cleared and disturbed to such an extent that the loosened detrital materials were now able to move downslope and downstream at a much faster rate than

The Formation of New Placer Gold Deposits

under normal conditions. This major increase in the quantity of easily erodible gravels and the accelerated rate of erosion have contributed significant amounts of additional precious metals for reconcentration into more recently formed placer gold deposits.

Over time, extended periods of heavy runoff have further accelerated the processes of natural erosion and have assisted in the creation of ideal conditions for the formation of new placer deposits. Long periods of medium-to-heavy snow and rainfall have preceded extensive and prolonged erosional cycles and heavy stream runoff, which again is conducive to a further increase in the rate of erosion. This is because the soils, sand and gravel that have become supersaturated with moisture become much more fluid in nature and the materials are more easily eroded than dry or compacted material. In those placer districts where the natural erosion of hillside gravels normally provides a fairly constant supply of new gold values each year, the quantities of gold available for concentration have increased substantially during times of heavy precipitation and increased runoff.

As time goes by, these massive amounts of loosened sand and gravel are continually moved, removed, washed, sorted and resorted by spring runoff, normal water flow and flash floods. The same natural processes that had acted to form the original gold placers have again been engaged in the gradual process of concentrating the placer gold into new deposits. The concentration of placer minerals has been accelerated, however, because of the increased number of gold particles available for sorting action.

On average, most of the precious metal particles available for concentration in new placer deposits were relatively small in size, because most of the larger nuggets were recovered by the old-time miners. As a result, most of the newly formed placers will have a much more even distribution of gold values and will not necessarily be concentrated on or near bedrock. The deposition of smaller pockets and pay streaks located at various levels above bedrock would be much more likely, particularly in those areas where large boulders are commonplace.

Ideal conditions for the deposition of new placers would also be present in locations where there is a sudden change to a lower streambed gradient and/or there is a sharp reduction in the rate of water flow. Because of the large quantities of fine and float gold present in the materials, the periodic formation (and destruction) of "skim bars" on the surface of the inside bends in rivers and streams would also be likely. These deposits are thin-bedded layers of sand, silt and mud that form on the surface of fairly permanent gravel bars following periods of heavy runoff, and they are commonly destroyed by subsequent heavy water flows.

It is also quite likely that the erosion and reconcentration of the massive volumes of disturbed gravels and tailings will have created a number of large, low-grade gold placer deposits in some of the major river drainages below major placer mining districts. In addition, the mining of many placer deposits that were of marginal grade in the past could now be profitable, either because of a secondary enrichment through the addition of gold values over the intervening years or because of the net increase in the average gold prices, or both. Gravel deposits with an extremely low average yield per cubic yard or ton can currently be mined profitably.

The influx of new gold particles eroded from hillsides and ravines into stream and river channels also can provide small, new concentrations of the larger gold particles in cracks, crevices and riffles in bedrock that were exposed by previous mining activity or where the channel has been scoured to bedrock by accelerated erosion. These concentrations of gold particles are usually fairly spotty and erratic, and they almost always contain only relatively small quantities of gold. Even though small and localized, these locations sometimes contain richer gravels and large nuggets. Some of the historic placer mining districts provide an annual "gold crop" that miners can harvest by crevicing and sniping activity or by using small suction dredges.

Recent advances in technology and greatly improved mining methods and equipment now make it possible to mine and process a much larger volume of lower-grade gravels at a fairly low cost and a much higher degree of efficiency in recovery. New methods for exploration and sampling are also available, which can enable the modern miner to more accurately define the overall extent and average value of placer gold deposits.

Another important consideration today is the potential value of the other heavy placer minerals contained in the gravels that might be mined as byproduct or co-product minerals. The heavier placer minerals will be reconcentrated in the new placers along with the precious metals, but to a much greater extent because none of these minerals were removed during the original mining activity. The prices for some minerals have increased significantly over the years and it is now possible that the value of these other minerals may exceed the value of the precious metals contained in some placer deposits. Because they had little or no value at the time, these minerals were largely ignored by the early miners.

Considering all the factors involved, the recent concentration of placer gold into economically feasible deposits seems very likely, and it could be even more so if the full mineral potential is realized.

5
New Gold Crops After Heavy Runoff

IN THE MINING SEASON that follows a year of heavy storms and high water runoff, the chances of finding placer gold can increase significantly. Many recreational prospectors and small-scale miners have searched streams, rivers and hillside ravines for newly exposed placer gold after a winter of major wet storms and the subsequent heavy spring runoff. Quite a few of them have been rewarded handsomely for their efforts.

During periods of heavy runoff, tremendous quantities of river and stream gravels are moved and redeposited and there is an accelerated erosion of hillside gravels. Dry ravines that normally experience very slow gravel movement become raging torrents of water under these conditions, and some of them are scoured to bedrock. In the major placer gold districts, these conditions cause a massive influx of new gold-bearing gravels into the watershed systems as well as the rapid movement and resorting of gravels in rivers and streams. The situation provides ideal conditions for the formation and deposition of new gold placers. As a result, some of the previously mined streambeds may now contain small reconcentrated placers as well as new concentrations of gold particles available for sniping and crevicing by small-scale miners and prospectors.

The best places to look for placer gold include:
- Above, below and in the fissures, cracks and ridges in bedrock at the bottom of rivers and streams.
- On the downstream side of large boulders or other water obstructions in stream channels.
- In potholes and gravel bars below areas where the water has scoured the streambed to bedrock.
- In gravel bars that have formed on the inside bend of stream and river channels.
- In gravel bars deposited where water channel levels are below a fast-flowing section of the stream.
- In central dips in bedrock underlying the streambed.

- In "streaks" of thinner gravels that generally follow the bottom of stream channels.
- In any of the newly formed gravel bars following extensive flash flooding or heavy winter runoff.

Gold, because it is extremely heavy (specific gravity 19.3 when pure) and is resistant to chemical decomposition, is easily concentrated by the combined actions of gravity and water movement. Gold particles are normally deposited in the lower strata of gravels in rivers, streams and dry ravines, although smaller particles can sometimes be found near the surface. In addition, larger gold pieces (nuggets) are occasionally found near the gravel surface after times of heavy water runoff, as immediately following a fast snowmelt, cloudburst or periods of heavy rainfall.

Because of their size and weight, most larger gold nuggets move along the bottom of a stream channel and are subsequently concentrated in the natural bedrock riffles as well as in areas where the velocity of the water drops off sharply. Large pieces of gold will embed themselves into the first available crack, fissure or pothole in bedrock and will remain there unless subsequent erosion of the rock itself, or extremely turbulent waters, eventually exposes the metals to the force of the water flow. Even when subjected to secondary movement, the large particles will only be transported a relatively short distance to the next natural obstruction in the stream channel where they remain until the process is repeated.

This entrapment of large gold nuggets can occur in a large number and variety of places, especially when the streambed has been scoured by a surge of heavy water flow. However, short and intense torrents of fast-flowing water may exert enough force to displace even larger nuggets, dropping them at a point where the flow is either obstructed or it levels out and slows down for a short distance. As a result, nuggets are sometimes left high and dry on the shoreline, just behind large boulders, on the top of gravel bars, and occasionally on the top of relatively smooth portions of exposed bedrock.

In many areas where richer gold placers were mined in the past, a new crop of gold nuggets is often deposited during spring runoff. Any gold that was missed by the old-timers, and newly eroded nuggets from hillsides and tributary ravines, will quite often be moved and redeposited in new locations by any significant increase in the rate of erosion. This creates an opportunity for a type of prospecting called sniping or crevicing, where miners check and clean the new gold crop from natural riffles and cracks in the bedrock almost every year, particularly after heavy runoff.

The best places to search for placer gold are those in or near the known major placer mining districts. According to government statistics, significant placer gold production has been recorded in at least 18 states, including Alabama, Alaska, Arizona, California, Colorado, Georgia, Idaho, Montana, Nevada, New Mexico, North Carolina, Oregon, South Carolina, South Dakota, Utah, Virginia, Washington and Wyoming. Seven of these states produced well over one million troy ounces of placer gold. In order of rank as placer producers, they are California, Alaska, Montana, Idaho, Oregon, Nevada and Oregon.

Placer gold production has been recorded from over 250 counties in the 18 producing states listed above, which shows the very widespread distribution of placer gold deposits. Occurrences of placer gold have also been reported in Arkansas, New Hampshire, Maine, Minnesota, Missouri, New York, North Dakota, Tennessee, Texas, Vermont and Wisconsin. This brings the total of known placer gold occurrences to at least 29 states, with an additional identified potential in several other states. The platinum group metals have also been found in placers in several states, most notably Alaska, California, Minnesota, Montana, Oregon, Idaho, Nevada, North Carolina, Virginia and Wyoming.

Placer gold has been found in most of the areas in which gold occurs in lode (hardrock) deposits, particularly where heavy and extensive erosion has taken place. The metal has also been transported over long distances by ancient glacial action and movement in major rivers and streams—in some cases, for several hundred miles. Gold particles have also been freed by deep, secular decay of massive rock formations over millions of years, and they are sometimes concentrated in significant eluvial and residual (basically in-place) deposits—even though the original host rocks may have contained only minute quantities of the metal. Some fairly rich placers have been found where lode gold deposits have been disintegrated and eroded in place.

Small amounts of placer gold can always be found in those areas where it has been mined in the past, especially in the larger or richer known placer mining districts. These old placer mining areas are an excellent place to begin panning for a few "colors" (small flecks of gold), but it should be kept in mind that most of these districts have been covered by mining claims and permission to prospect must be obtained from the owners.

When prospecting in or near old gold-producing districts (including lode gold mines), good places to pan for the metal would include rivers, streams, canyons, dry washes and ravines, older bench and terrace gravels, alluvial fans, lake and ocean beaches, river deltas and floodplain gravels. Dry gravels can be sampled and transported to a stream or pond for panning, or water can be taken to the sample site.

The best places to find larger gold nuggets are also in or near the known major placer mining districts, especially in locations where the bedrock is either exposed or can be easily uncovered. Look for natural cracks, fissures, potholes, ridges, crevices or pronounced dips in the bedrock which lie directly in the path of the water flow during periods of both high and low water levels, particularly those which are almost perpendicular (at close to a right angle) to the direction of water flow. In addition, check out any natural obstruction that would create a "riffle" effect in the waters, such as behind larger rocks and boulders (downstream side), near-surface gravels on the inside bend of rivers or streams, in and around exposed tree roots and clumps of grasses, and any other similar obstruction.

Extremely small, flat and flaky pieces of "float gold" can be panned from the surface sand and gravels in most gold-bearing districts. Float gold is composed mainly of very small, flattened particles that can be supported by the surface tension of water (since gold will not "wet" in water), and they are often transported for hundreds of miles from their original source. These particles can be easily distinguished from mica (a form of "fool's gold" that is similar in color but very light in weight) by the fact that they will separate easily in the bottom of the gold pan along with the larger pieces of gold—while the mica will be washed out of the pan. However, if float gold is exposed to air it will be picked up on the surface of the water when it flows over the metal in the pan.

When sampling bench and terrace gravels or large gravel bars, the samples should be panned from the surface down to the bottom of each gravel structure, since the gold particles may be deposited in layers within the deposit or, more likely, towards the bottom. Large gravel deposits may also contain pay streaks of gold-bearing gravels located at different levels within the mass.

Many of the older placers have been subsequently altered to some extent by changing geological conditions operating over extended periods of time, and many of them have been reconcentrated into new placers of a different type or they have been partially or wholly obliterated by extensive erosion. Many older placers also have little or no relationship to the current terrain features (topography). They have been found on the tops and sides of hills and mountains, buried under a few feet to several hundred feet of lava or overburden, occurring as bands in cemented conglomerates, in low-lying areas where extensive erosion has disintegrated large volumes of rock in-place, in ancient beach deposits that have been elevated by uplift or dropped by subsidence, in channel-like layers and structures in terrace gravels and alluvial fans, and in localized concentrations within short stretches of dry canyons and ravines.

Searching for placer gold has tantalized many people throughout human history, and it still presents an exciting pastime for those who are willing to try their skill and luck in the hope of making a new discovery. When there has been heavy water runoff in the spring preceding the placer mining season, the potential for placer gold discoveries is greatly enhanced.

PART II

Placer Deposit Formation and Characteristics

Because of the relative ease with which the minerals can be extracted and the richness of some deposits, good stream placers are as eagerly sought today as they were in past times. Gold placers are often referred to as the "poor man's mine" because they can be worked profitably by a single miner with a minimum cash investment.

6
Placer Characteristics in the Great Basin and California

Editor's Note: The following content was extracted from the white paper presented by the author, Dave W. Parkhurst, at the 1983 Western Placer Mining Conference in Reno, Nevada. The original text has been slightly modified in places to adjust the oral presentation format to print format. No substantive changes were made in the process.

THE MOST NOTABLE "CHARACTERISTIC" of placer deposits is that the greater percentage of them are *not* characteristic [that is, they are not characterized by a set of attributes that they all have in common]. Each deposit tends to be slightly unique, and many of them vary considerably in the types of materials deposited within a relatively localized area.

Several dictionaries define a placer as an "alluvial or glacial deposit of sand, gravel, etc., containing gold in particles large enough to be obtained by washing," or, "any place where deposits are washed for valuable minerals." This type of definition seems rather limited. Defining a placer as any localized concentration of the heavier and more durable minerals that has resulted from surface erosion might be more useful for this discussion. It covers a lot of ground.

Gold-bearing placers are the best known and most widely sought placer deposits for several obvious reasons. However, placer minerals include not only gold, silver and the platinum group metals but also such minerals as columbite, tantalite, cassiterite, chromite, gemstones, ilmenites, rutile, monazite, scheelite, and zircon. The list of placer minerals increases considerably if mineral deposits resulting from residual, or eroded-in-place, placer concentrations are included.

For the purpose of this discussion, gold-bearing placers will be used as examples for the general distribution and characteristics of the placer deposits in the Great Basin and California regions. With respect to the general placer classifications by type of mode or deposition, all of them, with some variations, are found within this general area. A few examples of each type of placer occurring in the region, with several references to specific localities, are mentioned below.

In California, the best-known placers are the river and stream placers, both recent and ancient, found in the "gold rush" areas of the Mother Lode. Most of these are characterized by very well-rounded and polished gravels which were deposited and redeposited over extremely long periods of time. The more recent placers are usually reconcentrations of older gravels from Tertiary and Quaternary stream channels, which are now buried under several hundred feet of volcanic rocks. (Isn't it convenient, by the way, to date placers within an approximate time span of 63 million years? The word "Tertiary" is often used instead of saying, "I don't know *how* old it is.")

There are also huge placer deposits in river deltas and terrace gravels, most of which have been worked by hydraulic methods or large dredges. Most of the gold particles in these locations are well flattened and vary in size from very small to extremely fine.

Beach placers have been found along the coast of California from just south of San Francisco to northern Oregon. Almost all of these are characterized by fine sands containing smaller, finely polished gold particles. Many of these areas were worked by hand on a small scale from the 1850s until the late 1940s and will probably be dredged on a large scale in the future.

Eluvial placers have been found in several areas of the Mother Lode district and elsewhere. They are characterized by very rough and coarse gold particles that are contained in angular and subangular gravels and rocks. Most of these areas were later developed into lode mining districts.

Several desert placers have been found and worked in the Mojave area of southern California and elsewhere in the state, some of which were worked on a small scale by Mexicans well before the days of the California Gold Rush. Most of these placers are characterized by angular and subangular gravels containing very coarse gold particles, excepting the older stream gravel deposits where the gravels and gold particles are more rounded and polished.

More recent glacial stream placers, or gravels transitional from moraines, have been found and worked in the Sierra Nevada. The older moraines were sometimes eroded and concentrated in the Tertiary and Quaternary stream channels. These gravels are usually characterized by the presence of glacial till and silt, well-rounded pebbles and finely ground gold particles. Some attempts were made in the past to use hydraulic methods to mine intact moraine gravels, but the use of these methods has since been halted by law.

In Nevada, typical stream placers have been found throughout the state, but most of them were formed during the pre-uplift period of the Sierra and Basin ranges. Considerable evidence exists that the climate in Nevada was extremely wet at various times in the past. Some portions of these older gravels have been eroded and reconcentrated in alluvial fan gravels at the mouths of canyons in several locations. These locations include Spring Valley, American Canyon, and the Incline area to Washoe Lake.

More recently formed stream and terrace gravel deposits containing placer gold have been found in alluvial fan gravels throughout the state. These deposits are normally characterized by the presence of well-rounded and subangular stream gravels in fairly well-defined channels that flowed through mostly angular terrace gravels. The gold particles vary from well-rounded and polished to extremely coarse, depending upon the conditions existent at the time of deposition. Examples are found at Gold Canyon, Spring Valley and Osceola.

Several residual gold placers have been found in the state, and they are characterized by angular gravels containing very coarse gold particles. Two eluvial-residual placer combinations containing high gold values were found and worked in the Olinghouse and Placeritas districts. Whereas lode mines were also developed at Olinghouse, most of the original lodes were almost completely eroded away at Placeritas.

There were a number of very large lakes in Nevada at various times in the past, as evidenced by the distinct beach deposits found throughout the state. Several locations have been found where the beach sand and gravel were gold-bearing, but, to the author's knowledge, none of them have been mined economically to date.

The potential for the discovery of many older dry placer deposits on the North American continent, especially in the western United States, Canada, and Alaska, appears high. Most of the more economical placers will likely be fairly well hidden, or buried, as the old-timers were very thorough in their search for the precious metals. With use of the more sophisticated exploration techniques available today, it should now be possible to locate hitherto unsuspected and inaccessible dry placer deposits.

What follows is a description of the deposition and characteristics of desert and dry placers and the distribution of gold particles within the placers.

The distribution of gold particles in any placer deposit is dependent upon a variety of geological conditions existent at the time of deposition and also upon the subsequent alteration on those conditions. In desert placers, the distribution of gold particles is usually entirely different from that found in the average stream or river placer, excepting those deposits which were formed under similar geological conditions.

Most desert gravel deposits are composed of predominantly angular material, the major portion of which shows very little, if any, rounding or smoothing as a result of water transportation. Due to the absence of a fairly constant water flow and the angular nature of the gravel, the gold particles don't normally work their way downward to clay or bedrock as quickly as in a steadily flowing stream. Being suspended in the gravels at higher levels, the gold is more easily accessible to secondary movement by subsequent surges of water during spring runoff or following thunderstorms.

Because of the turbulence and tremendous cutting power associated with sudden surges of heavy water flow, the gold particles are redistributed throughout large volumes of gravel. Unless there are later, extended periods of a fairly constant stream flow, the heavy metallic particles have very little chance of being sorted and concentrated into confined placer deposits.

As a result, most of the desert placers tend to have a fairly even distribution of small- to medium-sized gold particles mixed within the mass of gravel. It is not unusual to find some of the larger and heavier gold near the surface or in the upper gravel layers, while the smaller particles are found at various lower levels beneath. This results from the fact that smaller pieces of gold can more easily work their way downward through angular gravels than can the larger and coarser gold particles.

Intermittent surges of water flow over dry gravels also tends to form multiple layers within the gravel mass. These layers are generally more spread out and composed of a greater mixture of different types of material than the stratified gravels normally found in stream channels. In most of the dry regions, these layers are commonly separated by narrow seams of clay or "caliche."

Depending upon the geological conditions existing at the time of deposition, some of these gravel layers may contain gold while others do not. In some instances, gold-bearing layers of gravel are separated by layers or barren gravel.

In desert regions, the water flow is of such a brief duration that the underlying gravels are not softened or saturated with moisture. As a result, the vibrating or shaking action of the gravel by water movement is limited to the surface of the gravel mass. As a consequence, the larger gold particles are deposited in narrow streaks of limited length and depth, while the smaller pieces are distributed throughout the surface gravels. The pay streaks tend to be very irregular in occurrence and can only be located by hit-or-miss prospecting. For this reason, the use of prospecting methods that apply well to areas having wetter climates are likely to be more of a hindrance than a help in attempting to find desert placers.

Because of the limited abrasion present in the formation of desert or dry placers, much of the material is not broken down sufficiently to release all of the gold particles. As a consequence, many desert placers contain locked-in values in the matrix rocks distributed throughout the deposit. Some of these placers contain sufficient quantities of the original ore minerals to warrant crushing the material for recovery of the locked-in values.

The very nature of gold deposition in dry placers creates conditions conducive to the formation of large, medium- to low-grade placer deposits. It is also conducive to the formation of hidden placer deposits, buried under a few feet to several hundred feet of overburden, as well as fairly rich localized pockets and streaks of gold-bearing gravel. Most of the remaining desert placers will likely be difficult to locate, and, once discovered, will require careful and intensive sampling to determine their value.

A subject of particular interest to every placer miner is the mining equipment used. Placer mining is affected by geological conditions and geographical environment: climate, topography, vegetation, type of materials in the deposit, and the ratio of the gold content per unit volume of gravel. All of these conditions have a direct bearing on the design, construction and operation of placer mining equipment. The equipment and methods that work well and with a fairly high degree of efficiency on a particular placer deposit would not necessarily perform nearly as well on another placer, even in the same general area. The proper methods and equipment must be designed and engineered for the particular placer deposit being investigated.

Placer mining on a scale exceeding hand methods is as much of a business venture as any other enterprise. The various factors that govern the success of a placer mining enterprise can be predetermined to a great extent. A great many failures in the placer mining business can be attributed to too much optimism and not enough practical engineering and business analysis.

The percentage of failures in the placer mining field can be greatly reduced through better dissemination of information, an increased awareness of the critical factors affecting the placer mining business, and a common-sense approach to operations in the field.

Overview of the Gold Bug placer from Trench #3 looking north, fall 1987.

7
Alluvial Placers

THE FORMATION OF ALLUVIAL (or stream) placer deposits by natural means involves processes very similar to those used by the mining industry in the mining, crushing and gravity concentration of mineralized ores. In the natural process, alluvial placers are formed as a result of the disintegration and weathering of mineralized veins and rock formations and the subsequent mechanical concentration of the heavier and more durable minerals in economical deposits by the combined actions of running water and gravity separation.

Placer minerals are released from their host rocks by such natural agents as the wind, frost, rain, flowing streams, temperature changes, chemical actions, growth of vegetation, and movement of the earth's crust. Working slowly throughout geological time, these agents gradually reduce hard rock to gravel, sand, silt and clay, thereby releasing the precious metals and other durable minerals contained in the original host rocks.

The disintegrated materials are gradually moved downslope to the nearest watercourse by the actions of wind, rain, snowmelt runoff, and gravity settling. Upon reaching a stream or river, the moving water sweeps the lighter materials away and the heavier placer minerals begin to settle towards the bottom of the streambed.

In order for mechanical concentration to take place, the placer minerals must possess three necessary properties: (1) a fairly high specific gravity (weight), (2) durability (malleability or hardness), and (3) chemical resistance to decomposition. Some of the placer minerals that have these properties include cassiterite, chromite, columbite, gemstones, gold, ilmenite, magnetite, monazite, the platinum group metals, rutile, tantalite, zircon and several other minerals.

The process of mechanical concentration in water involves a few principles based upon the relative differences in specific gravity, size and shape of the mineral particles, as affected by the relative velocity and movement of the water.

As a general rule, heavier minerals sink much more rapidly in water than lighter minerals of the same size. The difference in specific gravity between two minerals is much higher in water than it is in air. For example, the ratio of gold (specific gravity = 19.3) to quartz (sp. gr. = 2.6) is about 7.4 to 1 in air and 11.4 to 1 in water, due to the weight of the water (sp. gr. = 1.0). This is computed as follows: 19.3/2.6 = 7.4, and 19.3-1/2.6-1 = 18.3/1.6 = 11.4. As a result, the two minerals separate much more easily.

The settling rate of the particles in water is also affected by their surface area. When two particles have the same weight but different size, the smaller particle will sink more rapidly in water. In addition, the shape of the particles affects their rate of settling. A rounded particle will settle much more rapidly than a flat or coarse particle.

The ability of flowing water to move or transport a solid object depends upon the velocity of the water, and this ability varies according to the square of the velocity. For example, if the water velocity doubles, then the transporting power increases to about four times the original force. Conversely, if the water velocity is cut in half, the transporting power is reduced to the point where most of the material being moved is dropped. As a result, most placer minerals are dropped wherever the current flow decreases substantially.

Swirls and eddies in stream currents tend to raise the lighter materials from the bottom of the channel and allow the main current to wash them away. The water turbulence in streams and rivers also simulates the upward pulsations and vibrations of jigs and tables used for concentrating heavier minerals. This shaking action moves the lighter particles upward, where they are swept away by the current, thus enabling the gold and other heavy mineral particles scattered throughout the gravels to become concentrated on the bottom of the channel.

All of these factors operate together to mechanically separate the finer and lighter minerals from the coarse and heavy minerals. Over an extended period of time, the placer minerals can eventually become concentrated enough to form economical placer deposits. Abrupt changes in stream gradient, channel obstructions, meandering or widening of a stream can produce conditions that will enable the heavier and more durable placer minerals to drop, accumulate and concentrate.

For economical placers to form, a continuous supply of valuable placer minerals must be available for concentration. It follows therefore that the most favorable regions for economical placer formation are those containing mineralized rock formations that have undergone extensive decay and erosion and which also have an extended topographic relief (a fairly large drop in elevation over a long distance).

Those areas that have undergone a recent geological uplift, and newer canyons that have cut into older valleys with the consequent rewashing and reconcentration of the pre-uplift gravels, are even more favorable. As a general rule, the more times such reconcentration of older gravels takes place, the higher the relative degree of concentration obtained of the valuable placer minerals in the newly formed deposits.

Alluvial (or stream) placers are the most important type of economical placer deposits, and they have yielded the largest quantities of placer gold, platinum group metals, precious stones, tin and other minerals. Because of the relative ease with which the minerals can be extracted and the richness of some deposits, good stream placers are as eagerly sought today as they were in past times. Because they can be worked profitably by a single miner with a minimum cash investment, gold placers are often referred to as the "poor man's mine."

Economical placer deposits are sometimes formed solely through weathering of the host rocks with little movement of the gravels, but they are much more commonly formed as a result of water transportation, sorting and concentration. Gold placer deposits are most often found in areas in which lode gold deposits occur, even though some of the original sources of the gold particles could not have been mined profitably. In areas where vein deposits were fairly shallow in depth, erosion has sometimes been so complete that no trace of the veins can be found. Because gold is usually the most sought-after mineral, a description of the metal's deposition in placer concentrations is provided below.

The largest and heaviest gold particles (nuggets) are normally found a short distance from their original source, while the smaller and finer gold particles may have been moved a long distance from their point of origin. In areas that have a fairly high stream gradient, very little gold will be deposited in the gorges while extremely rich, shallow gravel bars can accumulate in the convex curves of the streambed or wherever the water flow is partially blocked by natural obstructions, and at those places where the water flow slows down in the more level stretches of the stream or river bed.

However, some of the heavier pieces of gold, or nuggets, may lodge in the lee of natural riffles and ridges or in cracks in the bedrock, even in locations where the volume and velocity of the water are fairly high. In certain localities, these natural gold traps are cleaned every year with subsequent gold deposition occurring during spring runoff in each of the following years.

Most gold-bearing streams will have barren stretches as well as richer spots, depending upon the conditions prevailing at the time the gravels were deposited. Stream gravels are seldom rich directly opposite the mouth of a tributary stream, even if both streams are gold-bearing, because of the increased volume and velocity of the water at the intersection. Very little gold is normally found in rapids and in the bottom of whirlpools because any gold that lodges in these places is usually thrown out or ground to a powder by the milling action of the gravels and boulders.

The richer gold placers are usually found on the inside curve of meandering streams, in pockets behind large boulders, in gravel bars that accumulate directly below major obstructions in swiftly flowing streams, and at the first point where the streambed widens or levels out immediately below a narrow steep section of the streambed.

Some of the richer placers are formed by the erosion of older placer deposits and the subsequent reconcentration of values. This secondary movement of the gold particles may be caused due to uplift or subsidence of the earth's crust to an increase in the rate of water flow during periods of very heavy runoff.

Sand and gravel in water become more or less mobile, which allows the smaller gold particles in the mass to slip downwards and settle on the more impervious layers (such as clay or bedrock) beneath the stream. The clays deposited by a stream become coherent and plastic and the finer gold becomes embedded in them. Clays also work their way under large boulders, where they are protected from the disintegrating action of the water. Very rich pockets of gold have been found in clay accumulations under large boulders, particularly those that are close to bedrock.

In addition, gold is quite often found in gulches and ravines that are presently dry or contain only intermittent water flow. Some stream deposits are found on the tops and sides of hills where they were left by streams that changed directions or disappeared entirely as the surface of the earth changed during crustal uplift or subsidence. Also, as streams cut deeper into the rock strata or the direction of flow changes, gravel bars are often left behind at higher elevations. These deposits are known as bench or terrace placers and they have been found to contain some very rich gold-bearing gravels.

The pay streaks (richer concentrations of gold) in placer gravels are sometimes at or near true bedrock, but they are also found quite often at various higher levels in the gravel mass due to the conditions existing at the time they were deposited.

The economical accumulation of placer gold in larger gravel deposits requires a long-continued adjustment between the rate of water flow, the amount of gold available for concentration, and the quantity of gravels moved by the stream or river. Lindgren noted the largest and richest placer deposits were formed in streams and rivers having a gradient of approximately 30 feet per mile. The gravels could not be too thick, had to be moving fairly slowly downstream, and had to be constantly water-soaked in order for the "jigging" process to occur. The presence or absence of these conditions provided rich pay streaks in some streams and sparsely distributed gold particles in others.

Gold particles have a definite tendency to occur in concentrated pay streaks that may be fairly narrow and rich. These pay streaks may not be located in the present channel of the streambed and, if they are, they may not necessarily be in the central portion of the channel. Most pay streaks are commonly irregular in outline, have branches and splits, and are sometimes discontinuous (absent in some spots). Even though a pay streak is present, some (and sometimes most) of the gold particles are usually scattered throughout the gravel mass. All gold-bearing streambeds should be thoroughly sampled to avoid missing any richer pay streaks that could be there.

It should be noted that a large number of gold placers have very little or no relation to the present-day topography or terrain features. This is especially true for those gravels deposited during the Tertiary or early Quaternary geological time periods—the so-called "ancient river" placers. Many of these ancient streambeds are buried under several hundred feet of volcanic rocks or accumulated sediments.

It is highly likely that most of the major alluvial gold placers have already been discovered and worked, but it is also likely that fairly rich localized placers can still be found by the determined prospector.

8
Beach and Marine Placers

BEACH AND MARINE PLACERS are formed by processes that are similar, but not identical, to those involved in the formation of most alluvial placers. They are actually just another form of alluvial deposits, since they are formed by the combined actions of water movement, transportation and sorting of minerals. However, beach and marine deposits exhibit some unique characteristics that put them in a separate category.

Beach placers are not only deposited in areas that border the world's oceans but also on the shorelines of freshwater lakes. They have been formed on existing ocean beaches and lakeshores as well as on ancient lake and ocean shorelines. Because of the movement in the earth's crust over long periods of time, wide variations in the level of oceans and lakes, and changing climatic conditions, beach placer deposits can now be found in desert areas, beneath existing oceans and lakes, on mountain tops and hillsides, in elevated beaches above present shorelines, and in older cemented conglomerates and sandstones. As with most other types of placer deposits, the present location of beach and marine placers may bear little, if any, relation to the currently existing features of the surrounding terrain.

In order for economical placer deposits of any type to form, an adequate supply of placer minerals must be available for concentration. In the case of beach and marine deposits, the minerals may be provided by the transporting action of rivers and streams (which can move tremendous quantities of placer materials over long distances and deposit them on or near the shoreline of any body of water); the massive disintegration and erosion of mineralized rock formations located near oceans and lakes; a continuous process of chemical decomposition, abrasion and sorting of materials on a relatively stable shoreline over millions of years, and the decay and erosion of lode deposits located on the shoreline itself. Only those minerals that have a fairly high specific gravity and a resistance to chemical decomposition, and which possess sufficient hardness or malleability, will accumulate and be concentrated into economical beach placers.

The mechanical concentration of placer minerals in beach deposits is accomplished by several interacting processes. The wave and tidal action on shorelines causes an almost continuous movement of the surface sands and gravels, and the resulting abrasion between the rock and sand particles gradually grinds the materials down into a fine sand. The lighter and less resistant materials are washed away by the same repetitive water motion, while the heavier and more durable minerals are left behind. In addition, the less chemically resistant minerals are slowly solubilized in the liquid and are removed.

A combination of wave and tidal action and shore currents tends to shift the beach materials from one location to another and back again, exposing lighter materials for removal while sorting and concentrating the heavier and more durable minerals. Additional quantities of sand and gravel are also eroded from the surrounding area by rainfall and rivers and streams and provide a fairly constant source of new material to the cleaning, sorting and concentrating process. When the waters recede for any appreciable period of time, allowing the beach sands to dry, then the prevailing winds also blow some of the lighter materials away—forming sand dunes a short distance inland.

Acting together over long periods of time, these processes gradually create accumulations and concentrations of placer minerals in the beach sands and gravels, by the separation and retention of the valuable mineral particles contained in millions of tons of material and the removal of the lighter and valueless mineral particles.

As the heavier placer mineral particles accumulate, they are subjected to additional concentration by gravity settling. Particularly during heavy windstorms, as the waves of water impact the shoreline they produce a pounding and vibrating effect that shakes and loosens the placer material along the shoreline. This jigging action in the water-soaked sands tends to cause the heavier mineral particles to settle downward, thus displacing the lighter materials beneath and raising them toward the surface—where, in turn, the water washes them away. If they are not exposed by subsequent erosion, the heavier mineral particles continue to migrate downward until they encounter the first impervious layer of clay or bedrock beneath the shoreline. Over long periods of time, the mineral particles continue to collect and concentrate in a layer on top of the impervious strata, and can eventually form an economical placer deposit. These sorting and concentrating processes can further reduce a tremendous volume of sand and gravel into a relatively thin, high-grade layer that contains almost all of the heavier and more durable minerals originally distributed throughout the larger mass.

As a direct result of the processes outlined above, almost all beach and marine placers are characterized by thin-bedded and layered concentrations of placer minerals, which largely consist of small to extremely fine, smooth or fine-grained, mineral particles. The bulk of the materials comprising beach deposits consist of a very fine sand, which does not normally contain any appreciable amounts of larger gravels.

The subsequent geological uplift and/or subsidence of coastal regions, including the more ancient shorelines that are now located in inland areas, often displaced beach placer deposits by from a few feet to several thousand feet in elevation. As a result, economical beach placers may now be in submerged areas or at points several miles inland from the existing shoreline—even in areas where the present beaches have very little valuable mineral content. The elevated beaches can be distinguished from other types of placer deposits by several characteristics, particularly the uniform stratification (or layering) of the materials as opposed to the channel or irregular terrace gravel configuration typical of most alluvial stream gravels. In addition, an unusually large volume of clean and fine-grained sand will be present in the material, with very little associated clay or silt.

The most favorable locations for the formation of economical beach and marine placers are where extensive erosion of mineralized rocks has taken place over extremely long periods of time and the elevation of the shoreline has remained fairly stable. Delta areas, particularly those formed by gold-bearing rivers and streams that have flowed in the same general channel system for long periods, can be ideal sites for the formation of economical beach and marine placers.

Many of the more ancient beach placers were formed over a period of several million years, with the wave and tidal action continually working and reworking placer materials in the same general area. The long time periods involved provided the conditions necessary to reduce, sort and separate tremendous volumes of sand and gravel. As a general rule, the low-lying coastal plains and more recently formed shorelines do not provide the conditions necessary to form economical placers, because of the minimal erosion and sorting action taking place in these areas.

Lindgren, in *Mineral Deposits* (1933), has mentioned a narrow strip of beach sands about 200 feet wide, near Nome, Alaska, from which over $2 million in fine gold (at $20.67 per troy ounce) was recovered, and he also refers to two older elevated beach lines located further inland in the same general area that were mined to some extent. He further noted that beach gold was being recovered on the coasts of California and Oregon during the same general time period. More recently, a major effort was organized to dredge submerged gravels offshore from Nome. This placer deposit has been described as a river delta that has been reworked by ocean waves and tidal action. Another beach placer was dredged for platinum and gold values at Goodnews Bay, Alaska, for a number of years, and this deposit is currently being evaluated for additional mining operations.

A large number of economical beach and marine placer deposits have been discovered and mined throughout the world, and they have produced a significant quantity of gold, monazite, precious stones, tin, titanium, zircon, the platinum group metals, and several other mineral commodities. Beach sands are presently the major world source for monazite (a rare-earth phosphate), ilmenite (an iron-titanium oxide), and zircon (a zirconium silicate). Large quantities of placer gold and platinum have been recovered from beach placers in the past, and bucket-line dredging operations on a large scale are currently being considered in several locations.

9
Desert (Dry) Placers and Gold Deposition

DESERT, OR DRY, PLACERS may have originated in any of the seven basic types of placer deposits, depending upon the geological conditions existing at the time in which they were formed. This would include alluvial placer deposits (stream placers), hillside or eluvial placers (transitional movement), residual placers (eroded in-place), bajada placers (specific type of desert placer), glacial stream placers (from moraines), eolian placers (concentrated by wind action), and beach or marine placers (formed by wave action on lake or sea shores). Many of the original placer formations will have been partially or wholly altered by changing conditions over extended periods of time, and some of the deposits will have undergone erosion and reconcentration into placers of a different type.

The deposition of gold particles in all types of placer deposits is dependent upon a number of geological conditions existing at the time each deposit is formed, and also upon subsequent alteration caused by any significant change in those conditions. There are, however, some fairly basic characteristics that are typical of each type of placer deposit, and these characteristics can be used in the determination of where concentrations of gold are most likely to be found.

In desert placers, the deposition of gold particles is normally quite different from that which is found in the average stream or river placer, excepting those placers that formed during geological time periods when the climate was more conducive to the formation of stream placers. Most desert placers are generally composed of predominantly angular gravels, the bulk of which show very little of the rounding or smoothing that results from the abrasion created by continuous water transportation and sorting. Because of this lack of a relatively constant water flow and the overall roughness of the placer gravels, gold particles do not usually work their way downward to bedrock or clay layers nearly as rapidly as they would in a steadily flowing river or stream. Since the particles are suspended in the gravels at much higher levels, the gold is much more accessible to secondary movement by subsequent surges of high water runoff.

Because of the turbulence and tremendous cutting power associated with sudden surges of heavy water flows over dry gravels, most of the gold particles are frequently redistributed and redeposited throughout large volumes of thin-bedded gravel layers. Unless the gravels are subjected to subsequent, extended periods of fairly steady stream flows, the heavier metallic particles have very little opportunity to be sorted and concentrated into richer placer deposits. As a consequence, most desert placers tend to contain a fairly even distribution of very fine, small- and medium-sized gold particles mixed in the gravel mass.

It is also not uncommon to find larger and heavier pieces of gold near the ground surface, with most of the smaller particles being deposited at a much lower level. This condition results from the fact that smaller particles can work their way downward much more easily than the larger and coarser gold. However, the situation may be reversed when the larger gold particles were eroded first or during periods of maximum water flow and deeper erosion.

The gravel layers in desert placers are usually much more spread-out and composed of a greater mixture of materials (soils, sand, pebbles, large boulders and organic material) than those occurring in stratified gravels characteristic of stream deposition. In arid climates, these layers may be separated by narrow seams of clay or "caliche" (cemented gravels) that have formed during extended dry periods.

In most stream placers, the constant vibration of the water-saturated gravels by fairly continuous water movement tends to settle the heavier and more resistant metallic particles toward the bottom of the channel. This process is very similar to the action of a shaker table or movement of a gold pan. In desert areas, however, the intermittent surges of water flow are usually present for very short periods of time, such as immediately following a thunderstorm, cloudburst or rapid spring snowmelt. The flow of water is of such a brief duration that the underlying gravels are not softened or saturated with moisture. Because of this condition, the vibration or jigging action created by water movement is normally limited to the top layers in the gravel mass.

Under these conditions, the larger gold particles are normally deposited in fairly narrow streaks of limited length and depth, while the smaller particles are distributed throughout the surface gravels. Most of the "pay streaks" tend to be very irregular in occurrence and can usually only be discovered by hit-or-miss prospecting efforts.

Due to this irregular pattern of deposition, the use of prospecting techniques that apply to areas having a more constant stream flow are likely to be more of a hindrance than a help when prospecting desert placers.

The sporadic nature of gold deposition in desert placers also tends to create layers within layers, where some of the stratified gravels within a zone of deposition (usually defined by bands of clay or caliche) are gold-bearing while other strata in the same layer are not. Fairly narrow bands of gold-bearing gravels are sometimes sandwiched between barren bands of gravel. The sampling of any desert placer deposit should always be taken from the top to the bottom of the gravel mass and, if any gold is found, each individual gravel layer, seam, strata or channel should be sampled. Some of these thin layers have been found to be fairly rich in gold content, and they could be very easily missed by the unwary prospector.

Extremely fine gold will remain thoroughly mixed in the gravels and will also travel long distances on the surface because of its ability to "float" on the surface of water. Float gold will often be transported for many miles from its source, and it will be subsequently moved even further with each new surge of water. In addition, many of the smaller and more rounded gold particles will settle much more rapidly through angular gravels (due to their greatly reduced specific surface area) than will the larger and coarser pieces.

The nature of gold deposition in desert placers creates conditions normally associated with the formation of large, fairly low-grade placer deposits. It is also conducive to the formation of hidden placer deposits, which may be buried under a few feet to several hundred feet of overburden, as well as the deposition of fairly rich, localized pockets and streaks of gold-bearing gravels. Many of these desert placers will be somewhat difficult to find and involve very careful and extensive sampling to determine their value. However, it is also likely that many previously discovered desert placers could now be mined at a profit because of the increase in gold prices.

Editor's Note: A portion of an earlier, related *CMJ* article is included below as an extension to the content in the above article that pertains to the formation and characteristics of desert placers. The original article was published in the August 1987 *CMJ*. The complete article can be found in the archives at the website for publisher Pine Nut Press (www.pinenutpress.com).

PLACER DEPOSIT FORMATION AND CHARACTERISTICS

Excerpt from **"Desert (Dry) Placer Deposits,"** *August 1987 CMJ:*

Some of the gravel layers in dry placers may carry significant placer minerals while others do not, depending upon the geological time period in which the heavier placer minerals were eroded and the cutting power and duration of each surge of water flow. In some instances the mineral-bearing gravel layers are separated by layers of barren gravels, or the gravel mass may only carry values in the upper layers. In others the layering can be reversed or occur in some unusual configuration, especially where the placer minerals have been subjected to erosion and redeposition several times.

Most of the alluvial fan gravels deposited in arid climates consist of multiple channel systems that interlace bench and terrace gravels laid down during periods of heavy runoff or flash flooding. In most instances, there are larger numbers of individual channel structures contained within the fan gravels than are found in alluvial fans formed in wetter climates. This results from the lack of a fairly constant stream flow, which is necessary to form deeper and more well-defined channel structures. Most of the channels are usually narrow and shallow because they are in existence for only short periods of time, commonly being cut off and buried by the next surge of material washed in by flash flooding.

These conditions can produce an entire series of multiple channel systems within the fan gravels, with the fan itself being spread over a wide area both laterally and vertically. This, in turn, can produce a very irregular pattern in the bench and terrace gravels, which do not normally exhibit the uniform stratification typical of this type of gravel deposit. All of these factors contribute to a more erratic distribution of placer minerals throughout the mass of gravel, since there is very little chance for the concentration of values by the transporting and sorting action of water that occurs in wetter climates.

It should be mentioned, however, that many of the desert regions show geological evidence of an abundance of water during certain time periods. The presence of fossil shells, lacustrine deposits, well-rounded and polished pebbles, and ancient lake and seashore beach lines, none of which could have occurred under arid conditions, is proof that these regions once experienced wetter climates. Any placers that were formed during these "wet" time periods, providing that they have remained fairly intact, should exhibit characteristics that are more typical of river and stream placers.

In most desert gravels, however, the extremely limited actions of water transportation and sorting create very little abrasion between individual pieces of the eroded materials. As a consequence, most of the particles of placer minerals are very coarse and jagged, and a significant amount of the values sometimes remain encased in the original host rocks. Some desert placers contain large chunks and boulders of ore minerals that have not been broken down enough to release their contained values. This is especially so in many of the residual placer deposits.

The nature of mineral deposition in dry placers creates conditions that favor the formation of large, medium- to low-grade deposits of placer minerals. The conditions are also conducive to the occurrence of hidden placer deposits, which may be buried under from a few feet to several hundred feet of detrital materials. In addition, there is also potential for the formation of fairly rich, localized concentrations of placer values.

Russell Fields, director of the Nevada Department of Minerals in the late 1980s and into the 1990s, on a prospecting trip with Dave Parkhurst (who snapped this photo) in the Nevada desert c. late summer 1988. Russ later served as president of the Nevada Mining Association for a number of years. He retired in 2017 as president of the Mackay School of Earth Sciences and Engineering at UNR in Reno.

10
Glacial, Eolian, and Ancient Placer Deposits

GLACIAL STREAM PLACERS (or gravel deposits that are transitional from glacial moraines) are described here mainly for reference purposes since they have not usually been a significant source of placer minerals. However, with the recent rise in precious metal prices and the availability of more efficient mining methods and equipment, it is likely that a number of glacial placers could now be feasible to mine.

Glacial Placers. The tremendous erosive power of glaciers can be seen by observing the large, dish-shaped valleys that were carved out of solid rock by glacial periods in the past. In some areas the glaciers have dredged up huge mounds of detrital materials (called moraines) and, where these moraines contained sufficient quantities of valuable placer minerals, the streams resulting from ice-melt as the glaciers retreated have transported and sorted portions of the gravels. A large number of localized glacial stream placer deposits were formed by this process, some of which contained sufficient quantities of the precious metals to warrant placer mining.

Most of the materials contained in glacial stream placers are characterized by angular to subangular gravels that contain very finely ground placer minerals and large quantities of glacial silt (or "flour"). Depending upon the length of time that these streams flowed, some of the gravels may have been smoothed, rounded and polished to some extent—but the rounded pebbles are usually well mixed with more angular material. The slow movement and massive pressure exerted by glaciers almost always grinds the material beneath them to an extremely fine powder, while the face of the ice mass fractures disintegrates and scours the rock formations in its path.

The glacial till in front of the glacier is slowly accumulated and pushed into a mound as the glacier advances, and it is left in a ridge-like moraine when the glacier melts and retreats. Successive advances and retreats by the same glacier often leave a series of these moraines which, when subjected to further erosion, look like a series of small hills in the middle of glacial valleys. In some localities, there is a sufficient amount of precious metals in the glacial till to constitute a low-grade placer deposit. And, where the moraines have been eroded and sorted by stream action over long periods of time, some fairly rich localized placers have been deposited. The general formation of this type of placer is the same as for the normal river and stream (alluvial) placer deposits.

Eolian Placers. Eolian placers are also described here primarily for reference purposes, as only a few of them have been of economic significance. The general process of concentration involved, however, is often a significant factor in the formation of several other types of placers. Eolian placers are formed almost solely by wind action, the wind blowing away the lighter materials from a deposit as the rocks disintegrate.

Where high winds prevail for much of the time, this type of concentrating process can be particularly important. In addition, the erosive force represented by the abrasion of wind-borne particles of sand on rock formations should not be underestimated. In some of the desert areas noted for frequent high-velocity dust storms, for example, large portions of the mountains have been worn away by this force acting over extremely long periods of time. In some of the more stable areas of the earth's crust, wind action over a period of several million years has moved a tremendous amount of material.

The erosive power of high winds is manifested in several ways: (1) the wind blows away surface materials and exposes the underlying rocks, thereby subjecting them to more rapid disintegration by the effects of temperature (expansion and contraction), moisture, frost, chemical decay, etc.; (2) the airborne dust particles act as an abrasive and aid in the disintegration of other surface materials, and (3) the removal of the lighter materials gradually concentrates the heavier and more durable minerals that are left behind. If this process is allowed to continue in areas where crustal stability is the norm, and the materials are not removed by other erosional forces, an economic placer deposit can result.

Both Lindgren (1933) and Vanderburg (1936) refer to several economic eolian placers that were worked for a time in Western Australia. Lindgren noted that this type of placer can only be formed in extremely dry climates, where a long period of subaerial decay has broken down mineral-bearing formations and lodes, disintegrating the material sufficiently for the windstorms to act as an effective force in placer concentration. However, other types of placer deposits often undergo a certain amount of surface concentration by the effects of wind action, and this sometimes forms an economic surface placer on top of an otherwise low-grade placer deposit.

Ancient Placers. Placer deposits that were formed in earlier geological time periods are commonly referred to as ancient placers. This category would, of course, include all types of placers formed prior to more recent times, most particularly the more famous Tertiary and Quaternary river channels found in California, Alaska, Australia, Canada and South Africa. These older gravel deposits and their subsequent erosion and reconcentration into newer placers have been primarily responsible for many of the major gold rushes throughout the world in the past.

In a few regions, entire mountain ranges were eroded away and the resulting alluvium gradually disintegrated by the constant movement and abrasion of gravels in ancient river systems over millions of years. Some of the rivers were in existence for such long periods of time that "white channels" were formed, where only the hardest, most chemically resistant and durable constituents were left remaining—primarily quartz, chert, gold and the platinum group metals. The less-resistant materials were either ground to powder and washed away to be deposited as sand, silt and clay or broken down chemically and solubilized. The transportation, disintegration and sorting of this tremendous volume of rock has resulted in some of the richest placer concentrations known.

Many of the older river systems were very extensive, and they deposited a considerable volume of placer materials in multiple-channel structures, bench and terrace gravels, and floodplain and delta deposits. Following their initial deposition, stream overloading, lava flows, volcanic ash falls, and regional subsidence eventually buried many of these gravel deposits to depths of tens to hundreds of feet. In California, the ancient river systems were first buried and then uplifted several thousand feet by crustal movement and then exposed more recently by newly eroded canyons and ravines. In other regions, such as in Alaska and Australia, many of the placers remained submerged or buried, and the contained placer minerals were eventually extracted by drift mining to depths of several hundred feet.

Bateman (1942) refers to several rich gold-bearing channels at Ballarat, in Australia, which were drift-mined at depths of about 300 feet, and to similar buried gravels near Fairbanks, Alaska, which were mined to about the same depth. In addition, some of the older gravel deposits were subsequently buried and compressed into conglomerates or "cemented gravels," forming lode deposits that range in age from Cambrian to Recent. These altered and metamorphosed placer deposits have been mined for gold in South Dakota, New Zealand, South Africa and California.

Almost all of the material contained in the ancient river and stream deposits is characterized by rounded, smooth and polished pebbles and stones. The placer minerals are usually finely ground particles unless they are malleable native metals, in which case they are usually flattened and smoothed flakes, small rounded particles, or larger rounded lumps of metal called nuggets. The older white channels also usually contain significant amounts of flour and float gold particles.

A number of these ancient river channels have also been identified in semi-arid and desert regions, especially in areas with more recent mountain-building activity as a result of accelerated geological uplift. Some of these gravels have been found on the tops and sides of mountain ranges, having been elevated, in several cases, several thousand feet. It should be kept in mind that all other types of placers (i.e., residual, eluvial, beach and marine, etc.) were also formed in past geological time periods and, unless removed by subsequent erosion, many of them might still be fairly intact.

Many of the more recently formed river and stream placer deposits are actually reconcentrations of the placer minerals originally contained in older placer deposits. These reconcentrations were often further enriched by the flux of more recently liberated placer minerals that, in some areas, provide a relatively constant supply of mineral particles for the ongoing processes of sorting and deposition.

In summary, the more ancient gravel deposits have been an important source of placer minerals in the past, and they could become a significant economic resource in the future.

Editor's Note: Another article on this topic, titled "Prospecting for Ancient Gold Placers," can be found in the Dave W. Parkhurst Mining Writing Collection archives at the website www.pinenutpress. The article was originally published in the November 1991 CMJ and is an updated, expanded version of the "Ancient Placers" section of this article.

11
Residual and Eluvial Placers

RESIDUAL AND ELUVIAL PLACERS contain concentrations of placer minerals that have resulted from the disintegration of mineralized rocks within a fairly localized area. Residual placers are formed by the in-place decomposition and concentration of mineral-bearing rock formations, gravel deposits or ore bodies in relatively flat or gently rolling terrain. Eluvial placers result from the same processes acting upon materials that occur in sloping terrain.

The one basic distinction between the two types of deposits lies in the fact that eluvial placer materials are subjected to greater gravity movement and transitional creep because of the sloping surface and water runoff, both of which increase the rate of erosion. As a result, the higher concentrations of eluvial placer minerals are usually located near the bottom of the slope or in a thin-sheeted layer near bedrock that extends a short distance downslope from the source materials. Since residual placer minerals are normally eroded in-place, they are usually subjected to only vertical movement. As the rocks slowly disintegrate, the heavier and more durable mineral particles migrate downward to the first impermeable layer (such as clay) or bedrock.

The weathering process consists of both mechanical and chemical action, usually operating in combination to produce both disintegration and dissolution of the rocks and their contained minerals. Mechanical disintegration is accomplished by frost and freezing water, expansion and contraction of rocks during temperature changes, and the growth of vegetation. The physical fracturing of the rocks accelerates chemical decomposition by reducing the size of the particles, thereby creating a much larger surface area available for chemical action.

The more unstable minerals then undergo chemical decay, with the soluble portions generally being removed while the insoluble residues slowly accumulate. The chemical action actually creates some new minerals, a number of which may be stable under conditions existing near the ground surface, that also accumulate with the other more durable residual minerals. Stable minerals like gold and quartzite do not usually undergo any chemical change, so they are subjected to varying amounts of residual enrichment near the ground surface as they are slowly freed from the enclosing matrix rocks.

The agents of decomposition operating at or near the ground surface include water, oxygen, carbon dioxide, heat, acids, alkalis, vegetable and animal life, and a few of the soluble decomposition products of the disintegrating rocks. Hydration is essential to the process, so water is an important factor for both decomposition and disintegration to occur. The available oxygen permits oxidation of susceptible minerals, and carbon dioxide dissolved in water forms a powerful solvent (carbonic acid). The oxidation of sulfides creates both sulfates and sulfuric acid, both of which are active chemical agents, and heat accelerates most chemical reactions. Carbon dioxide and organic compounds that aid in the decomposition and dissolution of minerals are supplied by vegetation, and bacteria are also active in the biochemical processes of decomposition and precipitation.

Surficial weathering does not usually extend to a depth much greater than a few feet up to 30 to 40 feet, but some locations are known where this decay has reached to a depth of between 100 and 200 feet. Due to the percolating ground waters, sulfides have been known to oxidize to depths greater than 3,000 feet. In any event, deep and long-continued decomposition is necessary for the accumulation of significant quantities of residual minerals.

The effects of weathering vary according to the physical and chemical nature of the rocks as well as with the prevailing climatic conditions. Dry regions, which normally experience a wide variation in temperatures daily and in different time periods throughout the year, are subjected to mainly mechanical disintegration of the rocks in the region. The same is true in extremely cold climates, where frost and freezing water are the primary agents of decomposition while chemical reactions are slowed considerably. Chemical weathering is the most active in warmer and moist areas where the larger amounts of rainfall provides both moisture and nourishment for the lush plant life, which in turn yields organic and humic acids that promote accelerated chemical activity.

A separate type of residual and eluvial placer is formed in tropical regions where massive rock formations have undergone deep secular decay, producing large placer concentrations of minerals that resemble primary ore deposits. Since these deposits are formed over extremely long periods of time, the constituent minerals have often undergone considerable alteration in both their physical character and chemical composition. Large deposits of iron ore, manganese, nickel, bauxite, tin, gold, kyanite, barite, clay, phosphates and other minerals accumulate as residual concentrations in this type of deposit. They are found not only in tropical and subtropical climates where the current conditions are conducive to extensive rock decay, but also in other regions where tropical conditions existed in past ages. This type of mineral deposit often contains significant quantities of other interbedded placer minerals that can be extracted as byproducts.

On average, a certain amount of sorting and concentration of minerals is produced by the combined actions of wind and precipitation, but the residual concentration of valuable minerals is largely dependent upon their relative abundance in each unit volume of the decomposing host rocks. As a result, most of the richer placers form by the disintegration of higher-grade ores or the extensive decay of massive mineralized rock formations. Most of the lower-grade deposits form by decomposition of lower-grade ores, a superficial decay of mineralized rock formations, or the partial disintegration of higher-grade ore minerals. Because a minimal concentration of values is normally typical of residual and eluvial placers, most of them occur as thin-bedded, disseminated, or massive low-grade deposits.

Residual concentrations of valuable minerals are usually accumulated almost solely through the removal of undesirable constituents in the rocks through the processes of surficial weathering. Higher concentrations of minerals are normally produced by a decrease in the overall volume of the materials through chemical decomposition and dissolution. If these processes continue over sufficiently long periods of time, the valuable constituents may gradually accumulate until their relative purity and abundance produce an economically viable mineral deposit.

The conditions necessary for the formation of economic concentrations of residual minerals include the following: (1) rocks or ores containing valuable minerals must be present; (2) the undesirable minerals must be soluble while the desirable minerals are relatively insoluble; (3) the climatic conditions must be favorable to chemical decay and mechanical disintegration; (4) the slope cannot be too great or the valuable minerals will wash away as soon as they are freed from the host rocks, and (5) the earth's crust must be fairly stable at the site of accumulation for a long period of time. If all of these conditions are met, then the valuable residual minerals can accumulate in sufficient quantity to constitute an economic deposit, and the deposit will not be destroyed by erosion.

Residual placers are sometimes called "seam diggins'," since they have been found directly above several significant lode deposits. The "roots" of the vein structures or ore bodies are usually found directly beneath the concentrations of residual minerals, except in those instances where deeper secular decay has completely disintegrated the original host rocks. Because eluvial placers are commonly found directly below the ore deposits from which they originated, they are often called hillside placers. Various combinations of the two types of placers are commonly found together near the original source of mineralization, and they are often interspersed with other types of placers in the major mining districts.

Residual concentrations of placer values sometimes occur on top of older alluvial placer deposits, especially some of those that have been left as either terrace or bench gravels, or placers that have undergone crustal uplift following their original deposition. This type of reconcentration can produce richer surface concentrations of minerals on the top of lower-grade stream gravels. The reconcentration of surface gravels in sloping terrain is further enhanced by the action of wind, rainfall and water runoff. In some areas, these processes have produced richer surface gravels than the gravels found on bedrock.

The physical character of most residual and eluvial placer material is usually quite distinctive, excepting residual concentrations of alluvial gravels as noted above. The majority of residual gravels are very rough, jagged and angular, and they show very little evidence of the rounding and polishing that is typical of most stream placer gravels. Unless chemically altered, most of the metallic particles and mineral compounds will retain almost the same physical character that was present when they were originally deposited in the hardrock materials. Intact crystals are fairly common, as are rough and course metallic particles. The soluble salts can form encrustations on the rocks or, if the soil and gravel is fairly permeable, form layers of caliche, clays or encrustations lying near the prevailing water table or bedrock.

Due to the nature of the processes involved in their formation, most residual placers can be very difficult to recognize unless their presence is suspected or subsequent erosion has exposed some segments of the deposit. In flat or gently rolling terrain, the surface soils and vegetation tend to obscure any evidence of their existence. Some of the more sophisticated exploration techniques (such as geochemical soil and plant analyses, geophysical and geomagnetic surveys, and infrared photography) can be used to target potential placer concentrations as well as hidden (buried) lode deposits. Eluvial placers could also be hard to identify and might be detected by similar means, but they are much more likely to have been partially exposed by water runlets and shallow ravines that are common in areas with a sloping terrain.

Another typical characteristic of most residual placers is that varying degrees of material decomposition and disintegration are quite often found in different areas throughout the same deposit. The rocks that have been subjected to the greatest amount of weathering have generally disintegrated more completely, while those that were shielded from the full effects of erosional forces will break down much more slowly. This produces a significant difference between the relative quantities of "free" and "locked in" values in the horizontal layering within the gravels as well as from one location to another in the same mineralized area. Variations in the physical and chemical character of the host rocks produce the same effect, since some types of materials are much more susceptible to decomposition than others. Hard and durable rocks may be almost unaffected by weathering, so almost all of their contained values will remain in the original rocks.

Most of the residual and eluvial placers deposited during earlier geological time periods will have undergone extensive alteration. Geological uplift and subsidence, coupled with the effects of erosion and volcanic activity over millions of years, may have buried some of these deposits at depths up to several thousand feet while elevating others to the tops and sides of mountain ranges. Those deposits subjected to accelerated erosion will have been partially or wholly reconcentrated into different types of placers, or they may have been dispersed in large volumes of alluvial gravels. Some of the deposits have been compressed and cemented into conglomerates and metasedimentary rocks. Of course, this applies to other types of placers as well.

12
Buried Placer Deposits

EROSION, AND THE RESULTANT formation of placer concentrations, has been taking place for many millions of years. In some areas entire mountain ranges have been eroded away, and subsequent uplifted formations have already eroded significantly in the same general locations. Considering the numerous possible variations in placer deposition that have already been found, and the wide range of time in which they were formed, it seems very likely that only a small proportion of the total available placer deposits have been discovered to date.

Mineralized rocks are gradually broken down to sand, silt and gravel in a continuous process throughout geological time. Ancient gold-bearing conglomerates are being mined as lode deposits, as in South Africa, to depths of several thousand feet. The gold in these deposits was originally eroded and concentrated in placer formations that were later buried and compressed into sedimentary and metamorphic rocks. In other areas, gold-bearing conglomerates have been uplifted and exposed to further erosion, forming reconcentrations of the original gold particles in more recently formed placer deposits.

Streams and rivers have flowed during various time periods over millions of years in almost all localities above sea level throughout the world. Many of these stream channels formed placer concentrations that are still relatively intact. Others have been uplifted and subsequently eroded into new placer deposits, or dispersed. Several areas have been found in North America where multiple age groups of ancient stream channels overlie each other, the major portion of each age group remaining intact. Others have eroded in segments, enriching placers formed in each newer channel configuration.

Eluvial and residual placers, as well as beach and marine placers, have often undergone many alterations over extended periods of time. Quite a few of these placers have been subsequently reconcentrated in stream channels, though a large number of them could still remain nearly intact.

The main reason that most of these older placers have not yet been discovered is that they are partially or wholly buried by surface sedimentation or volcanic rocks. Subsidence of the ground surface in some localities, coupled with the ongoing erosional processes, may have buried some of these deposits from several feet to several thousand feet. More recent lava flows have also covered many placer deposits with various depths of volcanic rocks.

Unless these buried placers are partially or wholly exposed as a result of later uplift or recent erosion, they will remain hidden from those individuals using standard prospecting methods. There are, however, several newer, more sophisticated prospecting techniques available today that might aid in their detection. Several economical buried placers have been recently discovered through use of geochemical, geophysical and geomagnetic methods. Also proving useful in this area is infrared photography.

Since most placer concentrations contain large quantities of other heavy minerals, or "black sands," the more sensitive electronic magnetometers have proved fairly efficient at localizing channel configurations, especially where the placers are not covered by deeper alluvium or rock. Once the buried channels are traced, they must still be either drilled or excavated to complete definition of the deposit.

A major problem encountered in prospecting a buried placer is in the location of the areas of highest concentration of the placer minerals. The geological definition of the deposit must be done on a grid average of the drill or test pit results. Unless the average distribution of gold or other valuable mineral particles is fairly even throughout the deposit, it would be very easy to miss the areas of highest concentration.

If the placer deposit is located on or near bedrock, seismic methods can often be used to successfully determine the most likely "dip" areas for possible concentrations on the bedrock. Where the buried channels are resting upon clays, or "false bedrock," however, seismic methods usually cannot produce satisfactory results.

Geochemistry is now being used extensively to pinpoint possible mineral occurrences in both lode and placer areas. Plant, soil, rock and water samples are carefully analyzed to detect trace elements present in higher-than-normal quantities. This method is proving very useful in identifying potential mineralization in areas where no rock outcrops or placer materials are exposed on the surface.

Geomagnetic anomalies are often present in highly mineralized areas, but they are also caused by massive intrusives of several types of rock. In placer mining areas where the concentrations of heavy minerals are fairly high, these anomalies can sometimes be used to indicate the possible presence of buried placers. Generally, however, only the larger masses of mineralized gravels in sedimentary formations or extremely high local concentrations of heavy minerals will be sufficient to cause a noticeable shift in the earth's magnetic field.

Careful prospecting of the surface can also reveal the presence of buried placers, especially if erosion has partially exposed localized portions of the deposit. Since many of the more ancient placer deposits bear no relation to the current topography, their possible presence can be suspected in even the most unlikely places. Ancient stream channels have been found on the tops and sides of mountains, under sedimentary deposits in broad valleys and on plateaus, in desert areas as well as wetter climates, and in known lode gold mining areas as well as in areas where no gold deposits have been discovered.

In several areas, ancient gravel formations were partially or wholly eroded by subsequent stream action, reconcentrated into newer channel configurations, and then buried by later erosion or surface subsidence. Some of these can be located by tracing the erosion from partially exposed segments of the older gravels, or by locating "clip" areas resulting from more recent erosion into the secondary age-group channels.

Prospecting for buried placer deposits involves very careful analysis and observation of many slight indications. Much of the data indicating the possible presence of buried placers must be obtained by inference. A careful study of the geological history and mineralization in an area can often be very helpful in establishing potential exploration targets.

Since most of the known placer mining areas have been prospected intensively, the most likely areas for the discovery of economical buried placers are in the vast open lands that have received only superficial exploration efforts. The tremendous potential for the existence of large numbers of minable buried placer deposits would seem to warrant a special effort in prospecting for this type of deposit.

PART III

Gold, Silver, Platinum, and Diamonds

The earliest written record of placer gold mining is found in Egyptian hieroglyphics that date from about 3400 B.C. In addition, fairly small amounts of gold have been found in pre-dynastic Egyptian graves that have been dated to a period of time lying roughly 6,000 years in the past.

13
The Geology and Technology of Gold

SINCE BEFORE THE BEGINNING of recorded history, gold has been the most sought-after metal mankind has ever known. Prior to the advent of a "civilized" society, gold was regarded as an unusual and precious metal and was thought by many to have mystical and even religious significance. This was largely due to its unique color and workability and the fact that it was probably the first metal known to man, being one of the very few metals that occur in a native state.

Gold is a malleable, ductile, chemically inert, yellowish metal, with an atomic number of 79 and an atomic weight of 196.967. It occurs in its native state as a single stable isotope with a specific gravity of 19.3 and a melting point of 1,063 degrees Centigrade. Gold is highly reflective to infrared radiation and to most of the visible light spectrum, alloys readily with common metals, and has a high electrical and thermal conductivity. It is weighed by the troy system, in which 480 grains or 20 pennyweights equal one troy ounce. One troy ounce is the equivalent of 1.097 ounces avoirdupois or 31.104 grams.

The fineness of gold refers to the proportionate weight of pure gold in an alloy expressed in parts per thousand: 1000 fine gold is 100 percent pure. Commercially traded gold is usually 995 or higher. The "karat" is used to measure the purity of gold in jewelry; it is expressed in 24ths. So 24-karat (24k) gold is 1000 fine or pure gold, while 14k gold is 58.33 percent gold. White karat gold is usually composed of gold, copper, nickel, and zinc, although platinum and palladium are sometimes used.

Considered only for its beauty and permanence in the past, gold has emerged in the late twentieth century as an essential industrial metal. Electronic computers use the metal in many parts of their complex circuitry to ensure reliable operation; and men reached the moon using gold to shield their spacecraft and equipment from the sun's intense rays. Turbine blades in jet aircraft engines are bonded to their rotors with high-performance gold-brazing alloys. Many other uses for gold have made it indispensable to industrial technology. But gold still retains its appeal as a decorative metal, and even in industrialized nations most of the gold consumed is used in jewelry.

Gold occurs mainly as a native metal, alloyed with silver and other metals, and as tellurides. Other gold minerals are rare. It is commonly associated with the sulfides of iron, silver, arsenic, antimony, and copper. Weathering and erosion cause gold in the free or metallic form to be released from primary deposits and to accumulate as nuggets and grains in residual deposits or placer deposits.

Lode gold deposits are found in many kinds of rock. Hydrothermal deposits appear to be more common in acidic and intermediate igneous rocks, in silty laminated carbonate rocks, and in siliceous or aluminous sedimentary and metamorphic rocks. General types of lode deposits include hydrothermal, hydrothermal metamorphic, metamorphic, replacement (lode, massive, and disseminated), and cavity-filling (fissures, stockworks, saddle reefs, breccias and conglomerates).

Although the richest lode gold deposits have been small fissure vein types with quartz gangue (the so-called bonanzas), much larger quantities of gold have been mined from large medium-grade deposits. In South Africa, for example, gold ores presently average about 0.25 ounce per ton, although some mines have ore as rich as .065 ounce per ton. U.S. gold ores average about 0.2 ounce per ton or less. The advent of cyanide heap-leaching now makes it possible for rocks, with suitable properties, containing less than 0.03 ounce per ton to be considered ore.

A number of new techniques have come into use in the exploration for gold, but the miner's gold pan remains one of the simplest and most efficient low-cost means for finding visible gold and tracing its source. Neutron activation analysis and atomic absorption spectroscopy permit relatively fast and reliable gold analyses, and have had an important effect on the ability of field geologists to perform geochemical surveys in areas of little previous interest due to high sampling costs and absence of outcrops. Geochemical studies have established relationships between gold and other elements such as tellurium, mercury, arsenic and copper, which can be useful as tracers for the presence of gold. Also, metallogenic studies relating tectonic history and regional and local geology have given a better indication of where important gold deposits may be located.

Wire-line core drilling and improved percussion drilling methods have reduced cost and increased efficiency in exploration and development activity. Research in sampling through statistical analysis techniques that take into account the distribution and size of gold particles in a sample has shown ways to reduce sampling costs. Computer methods have been useful in defining the extent of pay zones in large ore bodies, such as those in South Africa, and have application to low-grade disseminated deposits.

Improvements in mining machinery, the introduction of new gold-leaching methods, and higher gold prices have combined to lower the economic cutoff limits for gold deposits, providing incentive for reevaluation work and exploration of new areas.

Much of the world's gold production has come from deep, narrow veins or thin-bedded layers called "reefs." Many of these have been difficult to mine because of high temperatures, humidity, and extreme rock pressures. About one-fourth of domestic U.S. production of gold is produced by the Homestake Mine, a long-active and deep underground mine. Most of the remainder comes from open-pit mines.

Placer mining was an important source of gold in the past, and could become so again. Dredging offers the most efficiency in large alluvial or marine deposits. Draglines or hydraulic methods are commonly used in smaller deposits with large boulders. Most placer mining requires large quantities of water, but dry placer operation is also possible using air-blowing equipment for separation. Generally, efficiency in gold recovery is lower in dry placer processing.

Milling technology for gold is highly developed, and normal mill recovery rates may range from 92 to 96 percent. Gold is recovered by cyanidation, amalgamation, flotation, gravity concentration, smelting, or a combination of these processes. It is also recovered from anode slimes that accumulate in electrolytic refining cells, used in refining such metals as copper, cobalt and nickel.

Much of the world's gold is in government vaults, largely immobilized by agreements between the major industrialized countries.

14
Gold: Its Character and Concentration

GOLD IS PRESENT IN ALMOST ALL of the earth's crustal rocks, but it usually occurs in only minute quantities unless it has undergone some form of concentration. The average background level of gold content in rocks is close to 0.006 parts per million (ppm), and it is present in sea waters at an average of 0.000012 ppm. As a result, economical deposits of the metal are only found where geological processes have acted to produce local concentrations that are hundreds of times higher than these background levels.

The metal is in group IB of the Periodic Table of the elements, along with its chemically similar associated elements, silver (Ag) and copper (Cu). Since gold has an atomic weight of 196.967 and a specific gravity of 19.32 (when pure), it is about twice as dense as silver and almost three times as heavy as iron. The melting point is relatively high (1,063 degrees C.), and the element possesses both high electrical and thermal conductivity.

Gold is also very malleable and ductile, and one troy ounce of pure metal can be beaten into a foil that would cover an area about 30 square meters in size. The pure metal is fairly soft, about five times softer than diamond and only twice as hard as the graphite in an average lead pencil.

The metal's most common occurrence in nature is as the native element (Au), which normally exists in the monovalent state (Au+) or as the trivalent ion (Au+++). Gold is normally chemically inert in air or water and is also unaffected by most acids. Unlike most base metals, it will not combine with sulfur. The chlorine group of elements (known as halogens) react easily with gold, as do some of the halogen-base and alkali-cyanide solutions. The metal will amalgamate readily with the liquid metal, mercury, and it commonly forms a solid-solution series (alloys) with silver and, to a much lesser extent, copper.

The primary human appeal presented by the precious metal lies in its distinctive yellow color, strong metallic luster, and resistance to corrosion. In addition, it occurs in nature in many beautiful crystalline, dendritic and arborescent forms. The structure of the crystal lattice is a closely packed, face-centered cube.

Gold and silver have very similar atomic dimensions, which explains their common association in nature. In fact, native gold is seldom found in its most pure form, as it is much more frequently alloyed with up to about 20 percent silver (on average). The relative fineness of gold (and other precious metals) is a measure of purity, which is expressed in parts per thousand. The purest natural forms of gold are most often found in mechanical concentrations (placer deposits) and in the oxidized zones of gold ore bodies (gossans), where the leaching-out of the alloyed silver has greatly improved the gold-to-silver ratio. The minute quantities of precious metals present in crustal rocks are most commonly concentrated by one or more of the geological processes described below.

Tremendous amounts of heat and pressure are generated within the earth's crust along the geologically active mountain belts and, as a result, the rock formations are progressively subjected to both chemical and mineralogical changes. This process is known as metamorphism. At greater depths, a partial melting of rock formations releases magma (or lava) and volatile fluids into the overlying rock structures. This process establishes a temperature and pressure gradient both on a regional scale and, locally, in an area surrounding the upwelling magmatic intrusions.

The resultant volatiles and fluids flush the adjoining rock formations that are close to the heat sources, and many metals (including gold) may be scavenged from these areas by the reactive solutions. The gold-bearing solutions that are migrating away from the heat source can subsequently precipitate their dissolved gold load when they encounter a chemically contrasting type of rock. In other cases, the precipitation of gold may result from a reduction in temperature and pressure when the circulating fluids enter open structural zones, such as cavities and fissures.

Geologists have observed that geological conditions existing during several general time periods in the past have been particularly conducive to the formation of certain mineral concentrations, such as the relationship between gold deposition and the volcanic or mixed volcanic-sedimentary sequences. The gold deposits of the Precambrian greenstone belts and the gold concentrations associated with the much younger Cordilleran volcanic events have been related to volcanic episodes that took place in the distant geological past. Although the overall complexity of the geology makes it difficult to determine an exact interpretation of the events, the fact that much of the gold mineralization can be proven to have taken place well after the host rocks were in place supports the epigenetic view.

The major shear zones developed in the greenstone belts resulted from both crustal movement and the injection of igneous materials, and these wide zones permitted a lateral and vertical migration of potential mineral-bearing solutions throughout the rock formations. Several rock types within the greenstone belts, such as banded iron formations, contain a much higher than average background quantity of gold (roughly 0.02 ppm).

It is believed that scavenging of gold values from these formations may have resulted in the large quantities required for economical concentrations. Large amounts of calcium, iron, potassium and silica were also liberated by the scavenging solutions, which led to the bulk alteration of the rock piles along linear zones and the eventual precipitation of arsenopyrite, carbonates, pyrite, quartz, and other metallic sulfides. In addition, many of the secondary gold concentrations have resulted from a modification of the primary volcanogenic occurrences. The combined processes of chemical and mechanical disintegration of less-resistant rock formations, the removal of the decomposed materials from the site, and the gradual accumulation and concentration of gold and other placer minerals, are still occurring throughout the world. It is evident that these processes were also operating in the distant geological past.

The concentrating actions of rivers and streams resulted in the formation of auriferous gravel deposits (placers) that were the targets of the nineteenth-century gold rushes in North America and Australia. In addition, the ancient fossil placer deposits (incorporated into quartz-pebble conglomerates) formed the largest of all the gold discoveries on the planet to date, the Witwatersrand goldfields in South Africa.

It is clear that a great number of both deep-seated and surficial processes have acted continuously throughout geological time, producing a large variety of gold deposits.

Author's Note: Most of the foregoing information was obtained from several articles published in the International Gold Mining Newsletter, produced by the Mining Journal (London, England)

15
Types of Gold Ore Deposits

IN ORDER TO PROVIDE a better understanding of the various processes involved in the formation of gold deposits, it is often desirable to examine known occurrences for the purpose of devising deposit "models" to enhance the prospects for successful field exploration activities and to assist in the development and mining of newly identified resources. It's very difficult to devise an accurate system for sorting each type of gold occurrence into a specific category because of the almost unique character of each individual ore deposit and the fact that a multitude of both deep-seated and surficial processes have acted together throughout geological time to form a large variety of gold deposits.

In other words, gold is where you find it and it may occur in unlikely host rocks and locations where you would least expect it to occur. But the use of deposit models and the recognition of geological conditions known to be favorable for the deposition of gold can often provide an edge in the search for new deposits. Although no entirely dependable system for categorizing all types of gold deposits has yet been devised, an imperfect system can still serve a useful purpose if it is used as a general guide.

Along this line of reasoning, R.W. Boyle (1979) has classified gold deposits into the following list of eight general types of occurrences on the basis of the general geology and geochemistry of the world's known ore deposits.

Gold-bearing skarn-type deposits. Although most often associated with copper, lead, zinc and tungsten deposits, these types of deposits sometimes serve as host rocks for both gold and silver. Several are being mined primarily for their precious metals content.

Skarn-type deposits characteristically occur in calcareous formations located at, or near, granitic contacts and usually contain high-temperature calcium-iron-magnesium silicate minerals that have resulted from thermal alteration of the carbonate minerals, such as limestone and dolomite. The Copper Canyon ore body located near Battle Mountain, Nevada (the Fortitude Mine) is a classic example of the gold-bearing skarn-type deposit.

Gold and silver veins, stockworks, pipes and silicified ore bodies in deformed volcanic formations. Deposits of this type are found throughout the world in rocks of all ages, but they are most commonly found in the Precambrian greenstones and younger volcanic formations. Quartz is usually the most important gangue material and, while the older deposits are typically formed in tight veins with intimately intergrown minerals, the younger deposits are usually vuggy and display several distinct phases of growth, and the ore bodies may be irregular and discontinuous in nature.

The older rock sequences of the greenstone terrains are generally highly altered in the immediate vicinity of mineralization; the younger Tertiary host rocks often show only regional lithological changes. A large number of Canadian gold mines, such as the Central Patricia mine near Crow River and the deposits located at Porcupine and Pickle Crow, provide good examples of this type of deposit.

Auriferous veins, sheeted zones and saddle reefs in folded and fractured sedimentary terrains, and replacement bodies in fractured and chemically favorable host rocks. Deposits of this type are found mostly in folded marine sedimentary sequences that have often undergone extensive metamorphic alteration, and most of the ore occurrences are distal to the associated granitic formations. The gangue mineral is predominantly quartz, which usually contains varying amounts of pyrite and arsenopyrite.

The alteration of the host rocks may be minimal and the sedimentary rocks often contain enriched quantities of both gold and pyrite. The calcareous sedimentary rocks usually host the replacement ore bodies. The Telfer Mine in Australia falls into this rather broad category of deposits.

Auriferous granitic rocks, including dikes, stockworks and pegmatites. Although most unmineralized rocks of these kinds typically average only about 0.003 parts per million (ppm) gold, they are often enriched with the metal when they have been structurally altered by post-consolidation faulting or shearing. These geological events usually precede the emplacement of gold-bearing quartz veins, pyritization and alteration.

Gold and silver veins, stockworks and silicified zones in mixed volcanic-sedimentary sequences that are deformed and intruded by granite. These deposits are characterized by several of the features that are similar to some of the previous deposit types listed, and their specific geochemistry is dependent upon the composition of the rocks that host the ore deposits. Interbedded volcanic and sedimentary piles usually predominate in the area, and the gold mineralization is usually spatially related to, but postdates, the granitic intrusions. The Renabie ore body and the deposits of the Kirkland Lake area in Ontario, Canada, are prime examples of this type of gold ore deposit

Disseminated and stockwork gold-silver deposits in igneous intrusive volcanic and sedimentary rocks. These are a highly variable type of ore deposit, since they are found in a wide range of host rocks. They are normally characterized by the formation of large, low-grade, high-tonnage ore bodies, in which extensive silification is a common feature. On average, the base metal content is normally not significant.

This type of ore deposit is typical of the younger Cordilleran-type geological setting, and it includes the Carlin, Cortez, Getchell, and Gold Acres ore deposits in Nevada. Examples of this type of deposit formed in Precambrian strata are the Camflo and Lamaque mines in Ontario, Canada.

Gold deposits in quartz-pebble conglomerates. This category covers the largest known resource for gold in the world, and this type of deposit has accounted for over 60 percent of the world's total gold production over the past 50 years or so. Because these deposits are largely composed of ancient placer accumulations, the gold tends to be distributed as fine grains intermixed throughout the conglomerate matrix. The matrix is usually composed of water-worn, rounded vein quartz and quartz pebbles compressed into a fine-grained quartz sericite mixture. The gold is sometimes accompanied by uranium and lesser amounts of thorium, the platinum group metals and, rarely, diamonds.

The conglomerates in the ore deposits of the Witwatersrand Basin in South Africa were formed in relatively thin layers within major sedimentary piles, some of which have been altered by volcanic sequences. Buckshot pyrite is normally prevalent as are bands of carbonaceous materials, and the structures have formed what are commonly referred to as conglomerate reefs.

Placer deposits. Much like the original form of the quartz-pebble conglomerate deposits, placer accumulations derive from primary gold ore occurrences through the combined actions of mechanical and chemical weathering processes in combination with a subsequent mechanical concentration of the heavier and more durable minerals, including gold. Detrital gold particles have been transported by gravity in talus materials (eluvial deposits), or by flowing water to accumulate in alluvial placer deposits. Some people also think that the transport of some gold in solution has led to precipitation, under favorable conditions, and the formation of larger nuggets in placer gravels.

River and stream gravels were an important source of gold during the gold rush periods in North America and Australia, and gold placers are still being worked extensively throughout the world. The larger percentage of Russian gold is reportedly extracted from this type of deposit.

16
Separating and Identifying Gold Particles

THE SEARCH FOR GOLD has greatly intensified in recent years, primarily because of the relatively large increase in prices and demand for the precious metals and also because of the availability of fairly inexpensive equipment and methods used for recovering the metals. As a result, a significant number of novice prospectors have joined the ranks of small miners and prospectors in their quest for the yellow metal.

There are two very important questions that the beginner should have the answers to before initiating any prospecting venture. The first question is, Where and how can gold be found? The second is, How can I separate and identify gold once I have discovered the metal? The following information is intended to provide an answer to the second question.

Gold's General Characteristics

Native gold is a chemically inert, ductile and malleable metal. It has a yellowish-to-silvery color and a specific gravity (weight) that falls roughly between 19.3 and 15.2, depending upon the quantity and weight of the other metals present. Gold is never found in an absolutely pure state; it is always alloyed (mixed) to some degree with one or more of several other metals, particularly silver.

According to Lindgren, the relative fineness of placer gold (in parts per thousand) varies from about 500 to 999. Some silver is always alloyed with native gold, and copper is occasionally present. Other metals are rarely prominent in combination with gold, with the possible exception of palladium in some areas. Lindgren also stated that while vein gold (in ores) may have a fineness of 997 to 999, it is the exception rather than the rule (pure gold is 1000 fine). It is much more common for gold's fineness in ores to range from near 500, which corresponds to electrum (about half gold and half silver), to about 800 to 850.

It has been found that placer gold in any district will usually be of a much higher purity than gold occurring in hardrock deposits, with the relative fineness of the placer gold increasing with the distance it has been transported in water and with the decreasing size of the particles. As a result, while California gold from vein deposits averages about 850 in fineness, the transported placer gold in the more ancient stream channels averages between 930 and 950 fineness. It has been shown that this increase in the relative purity of placer gold results from the solution of part of the silver contained in the alloy from the outer layer of the gold particles by the action of the surface waters acting through time and the overall distance the particles are moved.

Gold occurs in nature primarily as a native metal alloyed with silver and several other metals, and occasionally as tellurides. Other gold minerals where the metal is chemically combined with other elements are rare, but gold is often mixed within several other mineral compounds, particularly in iron pyrites and certain carbonaceous ores (primarily as disseminated microscopic particles). Disseminated gold particles can usually only be detected by assaying the ore.

Gold particles contained in quartz and most other host minerals can be separated fairly easily by crushing the rock with a mortar and pestle and then panning the resultant powder with water. Native gold is commonly found in close association with the sulfides of iron, silver, arsenic, antimony and copper.

The weathering and erosion of primary ore minerals causes gold in the free, or metallic, state to be released from the original host materials and to accumulate as nuggets and particles in several types of placer deposits. Streambed gravels that contain concentrations of gold particles are usually the most sought-after source of gold by most small miners and prospectors because of the relative ease with which the gold may be separated from the other alluvial materials.

Separation and Identification of Gold

The miner's gold pan is still the best and easiest item of equipment to use for the separation of gold particles from other materials, especially by the novice or beginner. Proficiency in the use of the gold pan can only come through practice and experience, but many invaluable pointers in the proper panning techniques can be learned from experienced placer miners. Once the basic gold panning technique has been mastered, the next step is to be able to identify the gold and other minerals that are most commonly concentrated on the bottom of the pan.

There are several other minerals that may be encountered while panning that may be mistaken for gold, particularly mica and iron pyrite (an iron sulfide commonly called fool's gold). Mica may have a shiny, bronze-yellow color, but it is easily recognized by its very light weight and flat, platy cleavage. Pyrite usually varies from a brass-yellow to a silver white metallic color, and it can be distinguished from gold by its lighter weight (specific gravity between 4.2 and 4.5) and the fact that it will shatter when struck by a rock or hammer. Both of these minerals are relatively common in areas where gold is found.

Native platinum group metals and (rarely) native silver are some other precious metals that will be concentrated together with placer gold. The presence of these metals is fairly easy to determine since they are almost as heavy as, or heavier than, gold and they will separate with the gold in a gold pan. All of these metals have a silver-white to grayish color and the particles are usually malleable, being flattened when struck by a hard object. Placer gold almost always has a light-to-heavy yellow color in placer particles, which distinguishes it from other precious metals.

Some of the other heavier minerals that concentrate with gold in the pan are magnetite and hematite (iron oxides that compose the bulk of the "black sands"), ilmenite (iron-titanium oxide), marcasite (iron sulfide), rutile (titanium oxide), scheelite (calcium tungstate), wolframite (iron-manganese tungstate), tourmaline (boron-aluminum silicate), chromite (iron-chromium oxide), cinnabar (mercury sulfide), and zircon (zirconium silicate). All of these minerals can be easily distinguished from gold by their color, granular or crystalline structure, and lighter weight.

If sufficient gold is present in a matrix rock or mineral such as quartzite or pyrite, the additional weight of the mineral will cause it to separate with the heavier mineral concentrate in a gold pan while other pieces of the same material will not. Any heavier-than-normal particles should be checked for precious metal content.

A magnifying glass should always be used to examine the concentrated minerals in a gold pan and also to examine the mineral particles contained in ores. The magnification can often distinguish the characteristics of smaller mineral particles that cannot be seen otherwise. Most of the minerals that might be mistaken for gold can be largely eliminated by careful observation of the physical characteristics and relative weight of such minerals.

Free gold in placer gravels can normally be distinguished by weight alone unless the particles are extremely fine or thin. Gold, with its distinctive color and malleability, can almost always be identified by careful observation after it has been separated by gravity concentration in a gold pan. Because the particles will flatten easily, will not shatter when struck by a hard object, and have the distinctive yellow color, simple testing can provide absolute proof of gold's presence.

In addition, most placer gold from alluvial gravels will be at least partially rounded or flattened by the abrasive action of the materials in moving water, giving the particles a distinctive "nuggety" look even in very fine particles. The only exception to this is when the gold particles are very rough and coarse because they have moved only a short distance from their original source or have just broken out of the host rock. These particles can be fairly rough and jagged but will still flatten easily with a hammer. Native gold can also be cut easily with a sharp knife, without shattering.

Very fine or very thin, flat gold particles will "float" on water, due to the surface tension of the water and the fact that water will not adhere to the surface of gold. When concentrating this type of gold particle in a pan, it is necessary that the gold remain constantly under the surface of the water. If the pan is tipped so air reaches the particles, they will be picked up on the water's surface. This floating action can easily deceive the beginner because the gold appears to be much lighter than it actually is. Some panners use a small amount of dish soap in the pan to reduce surface tension and prevent the loss of fine gold particles.

Gold particles from some areas are slightly magnetic, due to small amounts of iron mixed or alloyed with the yellow metal. Particles of the metal that have not settled to the bottom of the pan will also be picked up by a magnet because the magnetic lines of force pick up magnetic minerals below the pieces of gold and thereby lift them out of the pan with the other minerals. Using magnets to separate magnetic substances from concentrates in a pan should be done with extreme caution.

The identification of gold in the field will become easier with practice and experience. Most experienced prospectors and miners can identify gold with ease and can also identify most of the other minerals commonly associated with gold by using simple field tests.

However, there are a few unusual combinations of gold-bearing minerals that even the most experienced miner cannot identify with certainty. Any time there is a reasonable doubt as to the identity of a mineral, the mineral should be assayed. There are also a number of both simple and sophisticated field tests that can be used with a reasonable degree of accuracy.

John Mahan panning samples for gold content at the Utopian Mine c. early 1990s. John and four others, including Dave W. Parkhurst, were partners in the Utopian project.

17
Placer Gold Mining

PROSPECTING FOR GOLD and placer gold mining have always elicited visions of excitement, grandeur, adventure and romance in those who have searched for the precious yellow metal. In the past, the quest for golden riches led many people into joining some of the major gold rushes that were primarily responsible for the rapid settling of isolated regions throughout the world.

Historical research has shown that early man obtained gold by very crude methods and, as a result, was only able to recover the metal from richer and more visible sources. The first gold nuggets were most likely found in stream gravels where, in later times, large amounts of gravels were painstakingly picked over and sorted by hand methods for their remaining gold content.

The earliest written record of placer gold mining is found in Egyptian hieroglyphics that date from about 3400 B.C. In addition, fairly small amounts of gold have been found in pre-dynastic Egyptian graves that have been dated to a period of time lying roughly 6,000 years in the past. It has also been proven that ditches and sluices lined with grassy turf or sheep wool were used for recovering placer gold sometime before the first century B.C. More recently, it has been discovered that rough gold pans, rockers and riffle-type sluices were being used by medieval Europeans in the early years of the fifteenth century.

Over this period of time, a large variety of methods were devised to prospect for, mine and recover the metal from gold-bearing gravels and the richer gold ores. Several of the basic gold recovery methods that used gravity concentration were nearly perfected in the past, and they are still in use today. The miner's gold pan and the sluice box are two examples.

Placer mining operations are based on the separation of gold and other heavy metals or compounds from eroded sands and gravels, primarily by gravity separation of the heavier minerals from the lighter materials. However, the mining and processing methods utilized for this purpose vary widely and depend upon a number of factors, such as: average grade and volume of the gravel deposit, relative size and distribution of the mineral particles, type and shape of the particles, presence or absence of sticky soils or clays, size and distribution of boulders, depth and uniformity of the mineralized gravels, availability of water, climate, accessibility of the mining area, environmental requirements, cementation or compaction of the gravels, relative size of the mining operations, and several other relevant conditions.

The methods used for excavating and handling placer gravels are normally determined by the size and type of processing equipment employed at a particular mining operation, and the size and type of processing equipment to be used are determined by the physical character of the gravels and minerals in the deposit. Some of the various mining and processing methods used (both past and present) include the miner's gold pan, batea, rocker, dip-box, long tom, dry washer, sluice box, ground sluice, hydraulic mining, suction dredge, bucket-line dredge, and a large variety of gravel processing plants utilizing a combination of grizzlies, shaker screens, trommels, spiral concentrators, jigs, bowls, cones, shaker tables, sluices, nugget traps and other devices.

A multitude of processing and mining techniques are being used to recover placer minerals, with almost as many slight variations in methods and equipment design as there are individual miners and equipment manufacturers. Each of the basic devices and their general application or use is mentioned in the information that follows.

The miner's gold pan is usually a flat-bottomed, circular pan that is made of plastic or metal, measuring from about 12 to 16 inches in diameter at the top and about 2.5 inches in depth, and with the pan sides sloping about 40 degrees. When used properly, the gold pan will efficiently concentrate heavier minerals in the bottom of the pan. It is particularly effective as a gold prospecting device or for use in the recovery of gold values from richer gravels and concentrates. An experienced panner can wash from one-half to one cubic yard of gravel per day, excavating the material with a pick and shovel.

A batea is similar to the gold pan, but it is constructed of wood and has a conical or rounded bottom. This device was used extensively by Spanish and Mexican placer miners in the past, particularly in dry placer areas, but it is not commonly used today. The panner "winnowed" dry gravels by tossing them into the air and allowing the dust and soil to blow away, shaking the gold particles toward the bottom, and scraping off the lighter surface gravels by hand.

Gold rockers (sometimes called cradles) are normally a combination of a washing box and a removable screen to sieve off over-sized gravel, with a canvas or carpet under the screen to catch fine gold particles, a short sluice with riffles to recover the larger gold, and with cradle-type rockers on the bottom of the assembly to enable the operator to shake the device with a sidewise motion. The devices can process from one to three cubic yards of gravel per day, depending upon the size of the unit and the number of operators. Some rockers have been mechanized with power equipment, thereby increasing their capacity and throughput rate.

The dip-box is basically a portable sluice that measures from 6 to 8 feet in length, with the bottom covered with burlap, carpet or canvas and a strip of about one-quarter-inch screen. The gravel is shoveled into the upper end of the box (with the box set at an efficient sluicing angle) and washed through by hand methods or by the use of a hose. Used primarily where water is in short supply, the dip-box can process about the same quantity of gravel as the rocker in a day.

A long tom essentially consists of a short input box connected to an open washing box from 6 to 12 feet in length, with a perforated wood or metal plate (or angled screen) at the lower end to screen off larger material, which then feeds the smaller gravels into a short sluice that contains riffles. The individual boxes are set at an angle so that the gravels will wash from one box into the next lower unit, with a slope of between 1 to 1.5 inches per foot. Mercury is sometimes placed in the riffle box for amalgamation. Two men can process from 4 to 6 cubic yards of placer materials per day.

Dry-washing assemblies have been constructed in a large variety of designs and sizes. They basically separate gold from sand and gravel by pulsations of air through a porous medium, usually canvas. The screened gravels pass down an inclined riffle box that has canvas or similar fabric on the bottom. A bellows is located under the box, from which a series of short, strong pulsations of air are blown through the canvas under the riffles. The combined action of the air pulsations and the shaking action of the device tend to concentrate gold particles on the canvas behind the riffles. The unit can be operated either by hand or mechanized equipment and, depending upon the size and type of unit and number of operators, it can process between 3 and 20 cubic yards of gravel per day. The higher volumes of throughput are obtainable only if the gravels are completely dry and thoroughly pre-screened. Some of the newer varieties of this device utilize electrostatic separation of the materials.

The sluice box normally consists of a long, U-shaped box that contains mats, riffles, corrugations, expanded metal or similar materials which trap the gold particles as water washes gravels through the box. These devices have been built to sizes ranging from a few feet long to a mile or more in length, and from a few inches to several yards in width. If operating under ideal conditions, two operators shoveling by hand can process from 15 to 30 cubic yards per day using a fairly simple 12-foot-long sluice.

Ground sluicing activity is largely a thing of the past because of environmental considerations. Extremely large volumes of gravels were washed (or "boomed") by heavy surges of water directed through confined trenches dug to bedrock or lined with log riffles in the bottom. A similar (but more environmentally acceptable) device was the "ladder sluice." This device was constructed of logs or tree limbs to act as riffles and emplaced, securely anchored, in the bottom of a stream or river; it was intended to act as a trap for the gold particles that washed downstream during spring runoff or during flash flooding. It was usually "cleaned" of the collected gold values once or twice every year when the water level in the stream dropped.

Hydraulicking is also mainly a thing of the past, again due to the environmental problems created by its use. This method is similar to ground sluicing, with the exception that high-pressure jets of water were directed into gravel deposits by nozzles (called "monitors"), thereby washing enormous amounts of materials through ground sluices, ladder sluices or extremely large sluice boxes (sometimes several miles in length).

Suction dredging, one of the most popular methods of placer gold recovery, basically utilizes a water- or vacuum-powered suction hose to suck up river and stream gravels which are then passed through sluice boxes or other types of recovery equipment. These units are constructed in a large variety of sizes and types. Most dredges are designed to process underwater gravels, but they can be adapted to run bank gravels by using specially constructed ponds.

The bucket-line dredges were originally designed during the late 1800s to process large volumes of river, bench and terrace, or delta and floodplain gravel deposits, and they are still being used in some areas today. The unit utilizes a long boom that contains a continuous chain of bucket-type "teeth." The revolving buckets excavate underwater gravels and convey them through various types of concentrating equipment. The entire unit is usually constructed on a floating barge. Some of the modern bucket-line dredges can process tremendous volumes of gravel each day with an unusually high rate of recovery. The earlier dredges were much more inefficient.

The gravel processing plants represent a multitude of designs, sizes and types of processing equipment and methods of recovery—so many, in fact, that they cannot be adequately described in anything short of a fairly lengthy book on this topic alone. However, almost all of these plants utilize the gravity concentration method for recovering gold particles, and some have additional provisions for amalgamation or cyanidation. A number of separate devices are usually linked together in series to accomplish the highest degree of recovery and concentration possible for each particular placer mining operation.

The excavation of placer gravels is accomplished by various means that are mainly dependent upon the factors noted previously. Placer mining operations can vary from pick-and-shovel size to large operations that use large dredging or earth-moving equipment, and they can process from a few cubic yards to thousands of cubic yards per day.

18
The Silver Minerals

SILVER IS A DUCTILE, MALLEABLE, silver-white metallic element with a hardness of 2.5 to 3.0 and a specific gravity between 10.1 and 11.1. The metal has an atomic number of 47 and an atomic weight of 107.87, exhibits a high resistance to oxidation, and possesses a high thermal and electrical conductivity. As with the other precious metals, it is measured by the troy system of weights.

Among the main silver minerals of economic importance are native silver, argentite, cerargyrite, proustite, freibergite, stephanite, pyrargyrite and polybasite. Some of the other silver ores include pearceite, stromeyerite, hessite, electrum (alloyed gold and silver), embolite, bromyrite, iodyrite, dyscrasite, and argento jarosite. There are at least 55 recognized silver minerals and lead is the metal most commonly found associated with them. Two other ores that are of importance because they usually contain silver sulfide in chemical combination are tetrahedrite (copper-antimony sulfide) and tennantite (a copper-arsenic sulfide).

The primary silver minerals are usually found in association with intermediate felsic igneous rocks, such as andesite and rhyolite. The ores commonly occur in veins along fault fissures and shear zones, and they range from a few inches to several feet in width. Ores from which silver can be produced as a byproduct mineral are usually found in copper porphyries, massive bodies of sulfide mineralization, or in copper-lead-zinc vein deposits. Silver has been found in all of the western states, and most of the U.S. exploration for new ore deposits has been taking place in this part of the country.

The vast majority of silver production has come from replacement-type ore bodies or cavity fillings where the metal is a byproduct constituent, most commonly associated with gold ores or with the base metals such as lead, copper, zinc, antimony and cobalt.

A basic familiarity with the physical nature of the primary silver ores is an aid both in their identification as primary ore minerals and in detecting their presence in complex ores and disseminated mineral deposits. A brief description of several of the important ores, their distinguishing characteristics, and most notable occurrences follows.

Native silver occurs as finely disseminated metallic particles in mineralized veins of hydrothermal origin, and it is sometimes found in masses or "wire" form. The metal forms cubic or octahedral crystals when it is crystallized, has a silvery-white color (but tarnishes to dull gray), is ductile and malleable, exhibits no cleavage, and is fairly soft. Some notable occurrences are in Arizona, Colorado, Idaho, Montana, Nevada, Canada, Australia and Germany.

Argentite or "silver glance" is a silver sulfide that often forms cubic crystals, but the faces are usually too distorted and branching to be easily recognized. The mineral is commonly found in masses with a dark lead-gray color and metallic luster, which normally tarnishes to dull black. It has a shining streak, hardness of 2.0 to 2.5, and a specific gravity of 7.3. Another form of the mineral, called acanthite, produces monoclinic, prismatic crystals, has an iron-black color and has a black streak. Argentite is the most common primary ore of silver; it is found throughout the western U.S., Chile, Bolivia, Mexico, Peru, England, Germany and Czechoslovakia.

Cerargyrite or "horn silver" is a silver chloride-bromide mineral with a higher proportion of chlorine than bromine. It is a secondary mineral found in the oxidized zone of major silver deposits, especially in arid regions. The mineral is colorless when pure and fresh, and develops a greenish-gray tint with an increasing substitution of bromine. It becomes violet-brown to purple on exposure to light. The mineral has a white or gray shiny streak, occurs in cubic, massive crystals, has a hardness of 2.5, and has a specific gravity of about 6.5. Notable deposits have been found in Arizona, California, Colorado, Idaho, Nevada, New Mexico, Australia, England, France, Germany, Italy, Spain, Argentina, Bolivia, Chile, Mexico, Peru and the [former] Soviet Union.

Bromyrite (as above) is generally a mixture of silver bromide and silver chloride similar to cerargyrite but with a much higher proportion of bromine. The mineral forms cubic crystals in parallel groups, but it is usually massive, in crusts, or in waxy and horn-like masses or coatings, columnar or stalactitic. Although its color is usually gray, yellowish or greenish-brown, it is colorless when pure and fresh. The mineral has a hardness of 2.5 and specific gravity of about 6.5. Notable occurrences have been found in Arizona, Chile, Mexico, Australia, France, Germany and the [former] Soviet Union.

Proustite, a silver-arsenic sulfide mineral, forms hexagonal, prismatic crystals and also compact masses. Its color is scarlet-vermillion; it has an adamantine luster and a vermillion streak. It has good cleavage in one direction, hardness of 2.0–2.5, and specific gravity (SG) 5.6. It has been found in Colorado, Idaho, France and Germany.

Freibergite is a copper-silver antimonide-sulfide containing up to 18 percent silver. It is found in association with tetrahedrite in Germany and Idaho.

Stephanite is a silver-antimony sulfide mineral that forms orthorhombic, tabular or prismatic crystals, and also occurs in masses. Iron-black in color, it has a metallic luster and an iron black streak. The mineral has an imperfect cleavage, hardness of 2.0–2.5, and SG 6.25. Notable occurrences are in California, Nevada, Canada, England, Germany, and Czechoslovakia.

Pyrargyrite, a silver-antimony sulfide, forms hexagonal, prismatic crystals and in masses. Its color is a deep red with an adamantine luster and purplish red streak. The mineral has a conchoidal-to-uneven fracture, cleavage in one direction, hardness of 2.5 and SG 5.8. Notable deposits have been found in Colorado, Idaho and Germany.

Polybasite is another silver-antimony sulfide that forms monoclinic, tabular crystals and masses. It is iron-black in color, has a metallic luster, and has a black streak. The mineral fractures unevenly and has a poor cleavage, hardness between 2.0 and 3.0, and SG of 6.1. It occurs in Colorado, Montana, Nevada and Canada.

Electrum. This is a gold-silver alloy that approaches an even distribution of the two metals (argentiferous gold). The alloy forms cubic crystals but is usually massive, is malleable and ductile, and is yellowish-silver in color. The specific gravity varies from about 12.5 to 15.5, depending upon the ratio of gold to silver. The most notable occurrences have been found in Nevada.

Although there are many other silver ore minerals, those listed above are among the most important economically. Because of their importance as significant sources of silver, **two copper-bearing minerals** are also described below.

Tetrahedrite is a copper-antimony sulfide; it forms cubic, tetragonal crystals and masses. The color is flint-gray, iron-black or dull black, and the streak is black, brown or cherry-red. Its hardness is between 3.0 and 3.7 and it has a specific gravity between 4.8 and 5.1.

The mineral occurs widely throughout the West in hydrothermal vein deposits.

Tennantite, a copper-arsenic sulfide, forms cubic, tetragonal crystals and granular masses. Its color is from flint-gray to iron-black. It has a metallic luster and a black or brown streak. The mineral has no cleavage, hardness from 3.7 to 4.5, and SG from 4.6 to 5.0. The mineral is very widespread in hydrothermal vein deposits (as above).

It has been estimated that about two-thirds of the known world silver reserves and resources are contained in copper, lead or zinc deposits. The veins or silver minerals in which silver is the primary component have been estimated to contain the remaining one-third of the total known world reserves and resources. Revisions in these figures are likely, because the higher average gold and silver prices and more cost-efficient mining methods have steadily increased the reserve base for both gold and silver contained in massive, low-grade precious metals deposits.

19
The Mineralogy and Geology of Platinum Deposits

Author's Note: Until recently, very little emphasis has been placed on exploration and development of the platinum group metals in the United States, and it seems likely that these metals present a tremendous potential for significant new discoveries. In addition, several of the base metals associated with platinum occurrences, most notably nickel, cobalt and chromium, are also very much in demand. The U.S. is currently reliant upon foreign imports for most of its supplies of these critical and strategic metals.

THE PLATINUM GROUP of metals consists of six closely related elements and includes platinum, palladium, rhodium, ruthenium, iridium, and osmium. All of these metals commonly occur together in nature, and they are among the rarest of the metallic elements. Because of their rarity, unique physical and chemical properties, and relatively high per-unit price, they are classed as precious metals and, along with gold and silver, are measured and sold in terms of troy ounces.

Although platinum occurs in the native state (chemically uncombined) in nature, it is almost always alloyed with one or more of the other metals in the group and minor amounts of several other metals, such as iron or copper. Platinum and palladium together normally account for 80 to 90 weight-percent of the metals found in nature, followed by ruthenium, rhodium, iridium and osmium in order of decreasing quantity. The metals are often slightly magnetic (particularly when alloyed with iron), and some metallic particles will attract iron filings.

The platinum group metals are siderophile elements (related to iron), and they primarily occur as magmatic products in basic and ultrabasic rocks. Nearly all of the metals are now obtained from lode deposits in these two types of rocks, where prior to the 1930s almost all of the world's production came from placer deposits. The metals also occur in association with high-temperature copper ores, and palladium usually predominates in this type of ore deposit.

The type of platinum occurrences are generally classed according to the following categories:
- Placer deposits.

- Disseminated particles in peridotite and olivine gabbro (commonly associated with chromite).
- Constituents in magmatic deposits in basic rocks (with palladium and quite commonly associated with chalcopyrite and pyrrhotite).
- In quartz vein deposits (usually in small quantities and associated with palladium).
- In contact-metamorphic deposits.
- In copper deposits of several types (normally in trace amounts with palladium).
- In basic rocks and ultrabasic rocks.

In basic rocks these metals are commonly associated with nickel and copper minerals and are usually combined with one or more of such elements as antimony, arsenic, bismuth, sulfur or tellurium. In ultrabasic rocks, the metals occur as native alloys, discrete minerals and in solid solution in the crystalline lattices of other minerals such as forsterite, spinel and the various sulfides.

Platinum Ore Deposits

Certain specific types of mineral deposits have formed as integral portions of igneous rock masses, which shows that they have originated in their present form by processes of differentiation and cooling in molten magmas. Most of the minerals formed in this type of process have a fairly simple composition and they are relatively few in number. Some of the most notable of these minerals are magnetite, ilmenite, cassiterite, diamond, pyrrhotite, pentlandite, chalcopyrite, molybdenite, arsenopyrite, the spinel minerals, lollingite, corundum, and the platinum group metals and compounds. At some locations, the platinum metals deposited in this manner have been found to be extensive and fairly rich in concentration but, in general, these types of deposits have been of much less importance than those formed by aqueous solutions.

Some of the most prominent characteristics of this class of mineral deposits are that they are normally included as a portion of a body of massive igneous rocks and the crystals of the minerals formed in magmatic solution tend to remain either disseminated throughout the mass or are differentiated into narrow bands, layers or "reefs" within the structure. It is usually somewhat difficult to determine where the highest concentrations of valuable minerals occur within the rock formation, since most of the material within the structure looks much the same. As a result, economic concentrations of ore minerals in these types of rock formations can be easily overlooked. It is usually necessary to perform a careful microscopic examination of the rocks in each distinct band or layer and/or attempt to identify the presence of certain "tracer" elements or compounds in the rocks that might indicate the potential occurrence of valuable mineralization.

As noted previously, in basic rocks the platinum group metals occur primarily as disseminated minerals consisting of one or more of the metals within the group combined with one or more of such elements as sulfur, antimony, arsenic, bismuth or tellurium. They are most commonly associated with cobalt, copper or nickel sulfides, which may be the principal mining products as they are in Canada and Russia, or important co-products as they are in South Africa.

Again, as noted above, in ultrabasic rocks the platinum group occurs as discrete minerals, native alloys and in solid solution in crystal lattices of other minerals. The relative proportion of the six metals in these deposits, most particularly the proportion of platinum to palladium, varies considerably from one mineral deposit to another according to the composition of the parent magmas and the differences in the ore-forming processes. For example, most Canadian lode deposits contain about equal amounts of platinum and palladium, Russian deposits contain about half as much platinum as palladium, and South African deposits normally contain more than twice as much platinum as palladium.

The platinum group does not commonly occur in economic quantities in either alkalic or silicic rocks, but the metals are being recovered from ores deposited in these types of rocks in the final stages of copper refining. The metallic content in these types of ores is normally extremely low. Prior to the development of the Stillwater Complex in Montana, however, most of the platinum group metals recovered in the U.S. were extracted from copper ores mined from silicic or alkalic rock formations.

Platinum group occurrences have been reported from a variety of rock formations throughout the world and, for informational purposes, some of the most notable and/or unique deposits discovered in past years will be mentioned briefly in the following overview.

Platinum Occurrences

Several of the Russian platinum deposits have been reported in which the metals are intergrown with olivine, pyroxene and serpentine crystals and also as occurring in Paleozoic peridotite from the Solowioff Mountain, where the metals formed zonal crystals of magmatic origin that lie between grains of chromite.

Platinum has been found in dunite from the Tulameen River in British Columbia, Canada, and the serpentines from this area also contain minor quantities of the metals.

In the nickel deposits at Sudbury, Ontario, the platinum arsenide called sperrylite, in association with palladium arsenide, is found in small, silvery-white cubes intergrown with chalcopyrite and pyrrhotite. The same mineral was discovered in the Rambler Mine in Wyoming, where it is associated with the copper ores chalcopyrite and covellite. The Rambler ores formed in lenses of igneous origin in dioritic rock formations, and palladium was also present.

On Prince Edward Island near Ketchikan, Alaska, the Salt Chuck Mine carried disseminated bornite with some chalcocite, covellite and epidote in a dark-green pyroxenite. The ore deposit was mined for several years before it was discovered that the bornite contained appreciable amounts of palladium, and about 3,000 troy ounces of palladium were extracted in 1925. An analysis of the concentrates from this mine showed 40% copper, 1 ounce of gold, 5 ounces of silver and 3 ounces of palladium per ton.

Several unusual occurrences of platinum in quartz veins have been reported from the southern island of New Zealand, in Finland and also in Canada. A rare occurrence of platinum with wollastonite and grossularite in a contact-metamorphic deposit was discovered in Sumatra. In Brazil, a palladium-bearing gold deposit was found in limestones lying close to a contact with igneous rocks.

A gold-platinum-palladium ore deposit that was concentrated by processes of oxidation has been reported in a mesothermal lead-copper-gold replacement in limestone in the Yellow Pine District, located in southern Nevada. The material that contained the precious metals was plumbojarosite, which is a sulfate of iron, lead and bismuth. One assay from the area showed 3.46 ounces of gold, 6.4 ounces of silver, 0.70 ounce of platinum and 3.38 ounces of palladium per ton. The gold occurred in rough, black and spongy particles, and the palladium-platinum formed microscopic black grains. It is believed that the platinum metals were originally contained in the primary sulfides and were later concentrated by mineralized solutions in which they were suspended in colloidal form.

The largest known occurrence of the platinum group metals in the U.S. to date is in the Stillwater Complex of Montana. The platinum metals occur here in a thin band or reef in the banded zone of the complex. The reef averages 6 to 7 feet in thickness over a linear distance of about 26 miles. The ores grade from 0.5 to 2.0% in nickel-copper sulfides and are reported to contain an average of about 0.65 troy ounce of platinum and palladium per ton. The ore minerals consist of nickel-bearing, pyrrhotite with lesser amounts of pentlandite and chalcopyrite, and other associated minerals include olivine, bronzite, augite, chromite and anorthosite. It is believed that the lower ultramafic lineage magma body contained the platinum group metals, while the upper anorthosite lineage body contained the sulfur essential for the formation of the platinum group sulfides at the reef horizon.

Several different types of platinum deposits have been reported in the Transvaal of South Africa. One type of deposit is connected with the lower norite zone of the Bushveld Igneous Complex, which covers an area of over 150,000 square miles. This intrusive sheet is similar in many respects to the smaller intrusive at Sudbury, Ontario. A division into horizontal sheets of different composition, or pseudo-stratification, is characteristic of the norite zone, which is primarily composed of rocks that range from peridotite to anorthosite and norite. In the lower part of the norite zone there are segregations and veins of a peridotite rich in iron that contain olivine, phlogopite, diallage, hornblende, chromite and titaniferous magnetite. In some of these rock masses platinum occurs in amounts ranging from 0.1 to 7.0 troy ounces per ton. Flat-lying beds of chromite also occur in the norite zone, some of which contain appreciable quantities of platinum group metals.

Several of the South African platinum deposits are associated with small amounts of nickel, cobalt and iron sulfides, which are deposited in both norite and bronzitite. The platinum and palladium are commonly associated with the minerals pyrrhotite, chalcopyrite and pentlandite. The Merensky-horizon type reef forms several important low-grade platinum deposits that occur in a persistent flat sheet of diallage-norite that measures from 3 to 30 feet in thickness. These rocks contain platinum in amounts ranging from about 0.1 to 0.15 troy ounce per ton, and the formations extend for many miles.

In the Waterberg District of the central Transvaal, several brecciated quartz veins and composite lodes occupy faults in felsite or felsite tuff. The lode deposits vary from about 6 to 50 feet in thickness over a linear distance of about 2 miles. The quartz is banded and has a comb structure; it contains chalcedony, specularite, oxidized pyrite and chromiferous chlorite. Assays of up to 150 troy ounces of platinum per ton were obtained prior to 1926.

A platinum occurrence near Potgietersrust includes both magmatic and contact-metasomatic deposits in basal dolomite, where the platinum is closely associated with primary sulfides. One reef-type horizon averaged 0.3–0.5 troy ounces of platinum per ton over a 1,700-foot distance.

Some of the most important platinum discoveries in South Africa were made in the Lydenburg District. These deposits occur in a sheet of diallage-norite that is 150 feet thick, underlain by anorthosite-norite. Platinum occurs in the upper 50 feet of the sheet, and the ores assay an average between 0.1 and 0.2 troy ounce per ton.

Platinum in Placer Deposits

The platinum metals occurring in placer deposits normally form small, rounded and concretionary or knobby dark-grey particles. They are sometimes accompanied by bright, silver scales of iridosmine or ommatidium. These native alloys are very dense, chemically resistant and physically tough. Following the disintegration of their original host rocks, the metals are transported and concentrated in placers in the same manner as particulate gold, and they are often found in association with gold placers.

Many of the older placer deposits are characterized by the nearly complete absence of palladium, since this metal (and, to a much lesser extent, platinum, rhodium and ruthenium) is solubilized during formation of the placer. As a result, in many of the most ancient placers, such as the Witwatersrand gold placers in South Africa, the predominant remaining alloy is either osmiridium or iridosmine.

In the U.S., the platinum metals occur in small quantities together with placer gold in most of the gold-bearing districts in northern and central California as well as in southwestern Oregon, mostly in the areas where serpentine and peridotite are found. Platinum has also been found with gold in beach sands from southern Oregon, northern California and Alaska. Most recently, placer platinum was produced from a dredging operation located at Goodnews Bay, Alaska. Previous placer production has also been recorded from the Tulameen District in British Columbia, Canada.

Almost 95 percent of the world's platinum was formerly extracted from placers on the eastern slope of the Ural Mountains in Russia, where detrital platinum particles occurred in stream gravels that headed in certain Paleozoic intrusives of peridotite and pyroxenite (partially altered to serpentine). The platinum was associated with iridosmine, iridium, gold and chromite, and its fineness (parts per thousand) ranged between about 750 and 850, with the remainder being composed of iron, copper and other platinum metals in the alloy.

Past placer production was also recorded from platinum-bearing gravels in river beds and Tertiary conglomerates found in the Choco District of Colombia, in other areas of South America, and also from Ethiopia in Africa. In the 1800s, the occurrence of platinum in placers was also noted in: Novita, Santa Rita and Santa Lucia in Brazil; two locations in Borneo; the sands of the Rhine River in Europe; the Jocky River in Santo Domingo; from Rutherford County in North Carolina; La Francois Beauce, Canada; with gold near Point Orford on the coast of northern California; and from several locations in British Columbia, Canada. Additional isolated platinum occurrences were also reported from gold-bearing regions in the southeastern United States.

20
Prospecting for Diamonds

A MINERAL THAT POSSESSES great hardness and refractive power, the diamond is composed of relatively pure carbon crystals that have formed under conditions of extreme temperature and pressure. Diamonds are the most popular and expensive of all gem minerals and, when of poor quality, are used extensively in several industrial abrasive applications. The occurrence of this valuable mineral is worldwide, although its concentration in commercially feasible deposits is relatively rare.

Diamonds have been discovered in several locations throughout the United States, but they have only been recovered in significant quantities from Pike County, Arkansas, to date. According to several sources, the stones have also been found in Alaska, California, Colorado, Georgia, Idaho, Indiana, North Carolina, Ohio, Oregon, Virginia, Wisconsin and Wyoming. Most of the stones found in this country so far have come from alluvial placer gravels, except for a few small stones from the recent kimberlite discoveries along the border between Colorado and Wyoming.

According to Dana in 1884, diamonds have been reported from Franklin and Rutherford counties in North Carolina; from Hall County in Georgia; in Manchester County near Richmond, Virginia; in Cherokee Ravine (Butte County), at Forest Hill (El Dorado County), at Fiddletown (Amador County), and in Nevada County—all in California; from several unknown locations in Idaho, and on the beaches of southern Oregon.

Lindgren noted in 1933 that the diamonds found near Murfreesboro in Pike County, Arkansas, were found in decomposed peridotite and breccias of late Cretaceous Age, very similar to several occurrences in South Africa. He also reported that, from 1906 to 1920, a total of 5,300 diamonds averaging 0.4 carat each in weight had been recovered from this area. The largest stone found weighed in at 20.25 carats.

The metric carat is the unit of weight used to measure both gem and industrial diamonds. There are 2,268 carats in an English pound, and a kilogram contains 5,000 carats. The carat unit (0.2 gram) is further subdivided into points, with 100 points being equal to one carat.

Diamonds are normally formed with 8 atoms of carbon per molecule; the crystals are predominantly octahedral with the surfaces often curved and sometimes striated. The stones also occur in massive form as rounded or irregular grains, or pebbles, which usually have an internal radial structure. The colors vary considerably and include white to blush-white, several shades of yellow or brown, and sometimes orange, pink, red, mauve, green, blue or black—depending upon the quantity and type of impurities contained in the stones.

The stones have an adamantine to greasy luster, are brittle, and have a perfect octahedral cleavage. Diamonds have a hardness of 10.0 and a specific gravity of 3.5–3.53, and they are extremely resistant to attack by either acids or alkalis. They have been found in kimberlite pipes, in alluvial fans and stream placers, in wave-concentrated beach deposits, in conglomerate, and (rarely) in meteorites.

The only important host rock for diamonds is kimberlite, which has been described by Lindgren as a serpentinized rock derived from peridotite and occurring in two varieties: the so-called basaltic type and a lamprophyric variety containing phlogopite. Lindgren wrote that the mineral is porphyritic and it is usually greatly altered, and kimberlite normally contains varying amounts of olivine, apatite, perovskite, ilmenite, chromite, enstatite, garnet and diopside. Some other sources describe kimberlite as being a coarse-grained hypabyssal igneous rock of the ultrabasic clan, normally brecciated and altered, and containing olivine, serpentine, tremolite, bastite, bronzite, biotite, pyrope, chrome diopside and calcite.

The world's present primary source of diamonds is an altered, dark-green form of kimberlite found mainly in southern Africa. (In the past, the major supplies of the stones came from placer deposits.) Even though about a thousand occurrences of kimberlite have been reported from various locations throughout the world, diamonds have not been found in the vast majority of them.

The term "diamond pipe" has been coined to describe a deposit of kimberlite that occurs in a pipe that is large enough and sufficiently diamondiferous to be commercially minable. The shape and size of these formations depends upon the relative position of the planes of structural weakness in the rock stratas through which the molten materials have passed. The pipes have been found to be columnar, tabular or irregular in shape. Where mining operations have progressed to a sufficient depth, the diamond pipes are usually found to decrease in area and constrict into dike-like structures.

The raw diamonds may contain inclusions of many other minerals and many of the same minerals may have inclusions of diamonds. These accessory minerals that may be "mixed" within the diamond crystals include garnet, olivine, diopside, ilmenite, magnetite, rutile and phlogopite. It has been noted that each mine site has certain diamonds which are typical of that particular location only, while every mine also contains some diamonds which are indistinguishable from those found at other mines.

Roughly 40 percent of the diamonds recovered from primary sources are currently being mined from alluvial placer deposits. These stones are being mined from river and stream placers, recent and/or elevated beach deposits, and glacial gravel deposits. In any of these mining areas, the diamond-bearing placer gravels were originally deposited in ancient drainage systems that are now totally unrelated to the present geological and topographical conditions. In some instances, the diamonds have been transported for great distances from their original point of origin.

Prospecting for diamonds can be quite challenging, since the specific gravity of the stones does not differ greatly from that of most mineral materials found in placer gravels. The detection of a valuable mineral in which the ratio is from 15 to 30 million parts of waste material to one part of the sought-after mineral, as is usually the case of diamond deposits, is not very easy.

The usual accompanying minerals in placers are known as the "satellites" of diamonds, and they include ilmenite, diopside, magnetite, garnet, zircon, rutile, corundum, monazite, epidote, topaz, spinel, tourmaline, gorceizite and staurolite. Although the average content of these minerals in the original kimberlite is fairly low, their presence in an alluvial deposit can indicate the possibility that they were transported from a primary ore deposit. It has been found that no reliable correlation may be made between the relative amounts of these minerals present in a placer and the quantity in diamonds occurring in the deposit.

If an alluvial area is suspected for potential diamond content, holes are sometimes drilled to bedrock and the cuttings examined for the presence of satellite minerals. If the results are encouraging, the area is then evaluated by sinking several shafts or excavating trenches. If satellite minerals are still found in quantity, then a large area is excavated to bedrock and a careful examination is conducted to see whether diamonds are present.

If extreme care is used, diamonds can be separated from placer materials in a gold pan or sluice box, although they cannot be concentrated effectively. After the lighter sands and rocks have been washed out of the concentrating equipment, the remaining material must be closely examined and any potential diamonds picked out by hand. Because diamonds are only slightly heavier than quartz pebbles (about 2.6 SG), a prospector should examine the concentrate before most of the quartz has an opportunity to wash out of the equipment.

Prospecting for kimberlite pipes presents an entirely different set of conditions for exploration. Other than by identifying individual specimens of the rock itself, massive kimberlite deposits can also be detected by magnetic exploration methods and by aerial photography. In some instances, plant growth can be indicative of changes in the soil quality, and this type of observation has led to the discovery of a few kimberlite deposits.

Most of the world's major diamond finds were made by individuals who just happened to stumble upon them. When one considers the widespread occurrences of kimberlite and the fact that diamonds have been found in many locations, it seems likely that other major diamond finds will be made in the future.

PART IV

Other Metals and Minerals Worth Seeking

Because of the fairly frequent association of mercury with gold and silver deposits, areas with identified mercury mineralization should also be checked for the potential occurrence of precious metals.

21
Beryllium: The Space Age Metal

BERYLLIUM WAS FIRST DISCOVERED by Vauquelin in 1797 and was produced as an impure metallic powder by Wohler and Bussy in 1828. In 1916, the first significant quantity of beryllium metal in ingot form was produced in the U.S., and the Brush Laboratories Co. began domestic development of beryllium resources in the early 1920s. The first significant industrial use of the metal occurred in 1926, following the discovery that an alloy with improved mechanical properties could be formed by adding beryllium to copper.

Little growth in the use of beryllium metal or the oxide took place until after World War II. This was because of costly and difficult extraction methods and the inability to machine the brittle and coarse-grained cast beryllium metal which was produced prior to that time. The markets for a beryllium-copper alloy, however, continued to be developed.

Over the past 40 years, metallic beryllium's unique properties have led to intensive research efforts toward increasing the domestic supply of the metal and to develop powder metallurgical fabrication techniques to produce a ductile beryllium that can be machined, formed, and used for specialized structural and other applications in the defense, nuclear, and aerospace industries.

From its first use in the 1920s, beryllium has now become a necessary industrial metal. Because of its high strength, light weight, and high thermal conductivity, the metal has been used in a large number of industrial and defense applications in the metallic form, as beryllium-copper alloys, and in the oxide form. The U.S. is, at present, the only market economy nation currently producing beryllium products from the ores and concentrates.

Bertrandite, a hydrous beryllium silicate, is the primary ore source for domestically produced products, with imported beryl ores (beryllium-aluminum silicate) providing the remainder of supplies for domestic consumption. The principal world beryl producers are Brazil, China and the [former] Soviet Union.

The U.S. has become less dependent upon imported beryl ores since 1969 and had the potential to become totally self-sufficient in the production of beryllium ores at the time of this writing in 1985. Known domestic resources of beryllium consist primarily of bertrandite ores located at Spor Mountain, Utah, and beryl ores located in the coarse-zoned pegmatites of New England, South Dakota, and Colorado. Some currently subeconomic resources of fine-grained beryl also occur in the Carolina pegmatites and in the subeconomic bertrandite deposits located at Gold Hill, Utah, and the Seward Peninsula, Alaska.

Author's Note: The above information was extracted from U.S. Bureau of Mines Bulletin 675, "Mineral Facts and Problems" (1985). It updates and supplements information published previously in the article "The Geology and Technology of Beryllium" (Dec. 1982 CMJ).

22
Chromium and Chromite

THE U.S. BUREAU OF MINES has studied the known areas of chromite mineralization in the U.S. to determine the potential availability of domestic chromium production in the event that foreign supplies of the metal were suddenly restricted or cut off entirely. The study found that the total known chromium resources in this country are severely limited, and most of the known chromite deposits are not economically feasible to mine at this time. The investigation underlined the need for expanded exploration efforts in America, with the objective of locating new, higher-grade chromite deposits to reduce the U.S. import reliance for the metal.

Chromium is one of America's most important strategic and critical metals because of its wide range of applications in the metallurgical, chemical and refractory industries. The metal's use in iron, steel and nonferrous alloys increases both the hardness and the resistance to oxidation and corrosion, and no practical substitute has been found to date. Yet in 1989, the U.S. imported approximately 79% of its chromium supplies, mainly from South Africa, Turkey, Zimbabwe and Yugoslavia.

The critical nature of U.S. dependency on foreign sources of chromite was recently emphasized by South Africa's threat to cut off exports of the ores because of the trade embargo initiated by the U.S. in efforts to end apartheid. South Africa accounts for the majority of the world's chromite production, and the country contains about 89% of the world's total known chromium resources. The event placed new emphasis on the fact that a much greater effort is needed to discover and develop significant chromite resources in North America.

Although chromium is a constituent of many minerals, chromite (an iron-chromium oxide) is the only ore mineral of economic importance. The ore does not usually form as a pure iron and chromium oxide but normally contains small amounts of aluminum and magnesium. Crystals are rare but are octahedral when present. The mineral is usually formed in masses, grading from fine-granular to compact. The mineral's color varies from brownish-black to black. It has a brown streak, uneven fracture, and no cleavage. It has a hardness of about 5.5 and a specific gravity of around 4.5, depending upon the impurities present.

The chromic oxide content of pure chromite ore is about 67.9%, but the content of natural ores is seldom above 50%. Replacement of some iron and chromium by aluminum and magnesium results in the formation of several varieties of chromite ores, including alumoberesovite, alumochromite, beresolite and magnesiochromite. Varying quantities of these minerals are usually admixed with chromite, as well as some accompanying magnetite.

Chromite is a mineral of the spinel group and it commonly occurs as an accessory mineral in peridotites and serpentines. The ores are normally more or less mixed with the host rocks and form irregular masses or flat bodies along the contacts or within the intrusive rocks. Several of the studies in the past indicate that massive chromite deposits are primarily magmatic separations in peridotite magmas.

However, in South Africa chromite occurs in pyroxenite and in rocks containing bronzite and anorthite, rather than in peridotite. The ores are found in podiform, stratiform and alluvial deposits, with the podiform and stratiform types being of the present major mining importance. Podiform deposits are those in which the ore occurs in the host rocks as irregular pods or lenses, and stratiform deposits occur as layers that normally extend over long distances. Both types are primary ore deposits, while alluvial deposits are secondary concentrations resulting from the erosion of the original ore bodies.

In the United States, chromite has been found in California, North Carolina, Maryland, Oregon, Pennsylvania, Texas and Wyoming as minable deposits in the past, and more recent discoveries have been made in Alaska, Montana and Washington. Lindgren noted that many chromite ore deposits were found in the great Serpentine Dike of California, mainly in El Dorado and Placer counties, although total production from these ore deposits was small. Several other chromite deposits were mined in California, Maryland, North Carolina and Oregon. The largest currently known U.S. chromite deposit is located in the Stillwater Complex of Montana, but the chromium ores are of the high-iron variety (low-grade chromium).

Chromite is also associated with gold, platinum, ilmenite, rutile, magnetite and zircon in some placer deposits. When it has been present in sufficient quantities, chromite has sometimes been recovered as a coproduct or byproduct in certain placer mining operations.

Chromite ore deposits also occur in some types of ultramafic rocks where the host rocks are composed primarily of olivine and pyroxene minerals or alteration products derived from these minerals.

Stratiform chromite deposits normally form layers of up to several feet in thickness, are usually fairly uniform in composition, and have been found to extend over large areas. The Bushveld Igneous Complex in the Transvaal, the Great Dyke in Zimbabwe, and the Stillwater Complex in Montana contain deposits of this type. Most of the world's known chromium resources are contained in stratiform ore deposits.

The podiform-type ore deposits are characterized by irregular forms, such as lenticular, tabular or pod-shaped bodies. Ore concentrations can range from a few pounds to several million tons. Most of the current production from this type of ore deposit comes from ore bodies containing about 100,000 tons or more. Some typical podiform ore deposits occur in the Ural Mountains of Russia, Albania, Zimbabwe, the Pacific Coast region of the U.S., and on islands in the western Pacific Ocean—especially in the Philippines. Most of the podiform ore bodies are of the high-chromium varieties (richer in average chromium content), and they are also the only current sources of high-aluminum chromite.

Several of the known ultramafic complexes that contain stratiform chromite deposits also contain platinum-bearing zones. Magmatic differentiation of these types of rock formations can also produce layers that contain concentrations of nickel and cobalt sulfides. A significant concentration of any one of these metals can indicate the potential presence of several of the other metals in the same geological formation or in adjacent mineralized zones.

Although a portion of the known low-grade chromium resources in the U.S. may become economically minable in the future, these resources could not supply more than a small fraction of America's requirements. Under current conditions, U.S. demand for the remainder of this century is expected to be primarily dependent upon imports from foreign countries.

23
Cobalt: A Strategic Metal

COBALT IS CLASSIFIED AS a strategic and critical metal because of its essential defense-related applications and the high degree of U.S. import reliance on foreign sources of supply. Although the United States is the world's largest consumer of the metal, usually accounting for about one-third of total world consumption, this country has produced no cobalt since 1971 (as of the time of this writing in 1986).

Cobalt is a silvery-gray metal with an atomic number of 27 and an atomic weight of 58.94, with a melting point of 2,723 degrees F. and a density of 8.9 grams per cubic centimeter. When magnetized, cobalt retains magnetic properties up to 2,050 degrees F., which makes the metal unique in its applications as an essential alloy in permanent magnets.

The metal has been developed from its first use as a coloring additive several thousand years ago into an essential industrial metal, due to its diverse physical properties. In most of its applications, cobalt imparts such essential qualities as heat resistance, high strength, wear resistance, and superior magnetic properties. Some of the major uses include applications in jet engine parts, cutting tools, electrical devices, permanent magnets, catalysts, pigments and paints.

The average crustal abundance of cobalt has been estimated at 20 parts per million (ppm), with the greatest crustal concentrations of the metal occurring in mafic and ultramafic igneous rocks and smaller concentration at the silica-rich end of the differentiation series. About 70 minerals contain cobalt as an important constituent, but only 20 of these commonly occur in economic concentrations. Almost all of the current cobalt production is obtained as a byproduct in the treatment of copper and nickel ores.

Most of the world's current cobalt production comes from stratabound copper deposits, though significant quantities of the metal have been produced from laterites, magmatic sulfide deposits, and hydrothermal ore deposits. The only known significant occurrences of stratabound copper deposits are found in Zaire and Zambia; they account for over half of the world's supply of cobalt. These deposits contain ores that grade from 0.1 to 2% cobalt, and the mineralization extends over an area about 300 miles long and 20 miles wide. The economic ore bodies occur in Precambrian sediments and metasediments, and the cobalt-bearing ores contain bornite, chalcocite, chalcopyrite, linnaeite and carrolite as well as the secondary minerals heterogenite, sphaerocobaltite, and erythrite.

Most cobalt resources are only available as byproducts in the mining of more abundant elements occurring in the deposit, since the production of cobalt as a principal metal requires the existence of a relatively high-grade, extensive ore body. The largest profitable mining district for the metal in the U.S. was the Blackbird District of Lemhi County, Idaho, which is located about 20 miles southwest of Salmon. This area was one of the rare locations where cobalt was mined as the principal product, but the mine was closed in 1959. The Blackbird District is considered one of the most likely areas for the resumption of cobalt mining in the U.S., mostly because of the relatively high-grade ores (averaging 0.6% cobalt).

Domestic mining of cobalt as a byproduct metal has primarily taken place at two locations: from a copper-nickel-cobalt mine in Missouri that was closed in 1961, and from iron ore deposits that were mined until 1971 in Pennsylvania.

The total known U.S. cobalt resources are estimated to be about 1.4 million tons of the metal, with most of the deposits occurring in Minnesota, Missouri, Idaho, Montana, California, Oregon and Alaska. Though fairly large, almost all of the domestic resources occur in subeconomic concentrations that cannot be mined in the immediate future. Minnesota has the largest identified cobalt resource: approximately 250 million pounds in sulfides occurring near Ely. These are very low-grade deposits, however, that are unlikely to be mined except in case of an emergency.

Missouri has the second-largest cobalt resources in deposits located in the southeast Missouri lead district, and the third largest resource is located in the northern California laterite zone. Large resources have also been identified in Alaska, Oregon and Pennsylvania, in addition to those occurring in the Stillwater Complex of Montana and the Blackbird District of Idaho.

A major source of strategic concern is the quantity of cobalt that would be available in the event of a prolonged supply disruption from foreign sources. A number of policies designed to prevent disruptions in cobalt supplies have been suggested recently, including: (1) increased acquisition of cobalt for the national stockpile, (2) subsidies to encourage production from domestic mines, (3) an industrial or economic stockpile, (4) increased federal funding for research and development of cobalt substitutes, (5) an expanded access to public lands for the discovery and development of domestic cobalt mines, and (6) accelerated development of ocean mining methods to exploit the cobalt-bearing manganese nodules and crusts.

Speculative resources of cobalt in both the U.S. and throughout the world amount to several billion pounds of the metal, but mining is only likely if the associated metals, such as nickel, copper, iron, lead, zinc, and silver, are in much higher demand than they are now. It would appear that an intensive land-based exploration program for higher-grade cobalt resources would be called for, or that deep-sea mining technology should be developed on a priority basis.

Recent research has shown that deep-seabed nodules containing manganese, nickel, copper and cobalt occur over wide areas of the ocean floor, and deposits of probable economic significance have been defined by sampling, photography and remote-sensing by means of sonar images. Potential recoverable resources in the northwest equatorial Pacific, the most promising area found to date, have been estimated at about 2.3 billion tons of nodules containing over 9 billion pounds of cobalt. Analyses of nodule materials from the Pacific Ocean near Hawaii show them to contain 21.6% manganese, 0.9% nickel, 0.66% copper, and 0.26% cobalt.

In addition, there is the possibility of mining manganese crusts or "pavements" on the ocean floor. Exploration of the Blake Plateau and in the central Pacific indicates the presence of deposits containing very large quantities of cobalt.

The Blake Plateau is a large submarine terrace located off the southeastern Atlantic Coast of the U.S., and it has been estimated to contain roughly 14,000 square kilometers of ferromanganese crusts — amounting to about 500 million metric tons of material. The cobalt concentration in the crusts averages about 0.3% cobalt.

The Mid-Pacific Mountains and Line Islands area in the Central Pacific Basin contain cobalt concentrations in ferromanganese crusts ranging from around 0.4% to 1.2%, with thicknesses of up to 7 centimeters. In addition, the crusts contain significant concentrations of manganese (25%), nickel (0.5) and several other metals. The values of metal oxides per square meter are significantly higher in the crustal deposits than in comparable materials from the known deep-water nodules. Indirect evidence also suggests that similar cobalt-rich crusts may occur in relatively shallow waters close to the Hawaiian Islands, the Trust Territory of the Pacific Islands, and other U.S. territories and possessions in the Pacific.

The basic U.S. problem with cobalt lies in its major dependence on overseas sources for primary supplies of the metal, combined with the fact that the known higher-grade ore deposits are concentrated in only a few areas of the world. Even though no primary cobalt has been produced in the United States since 1971, higher prices may make domestic mining profitable again in the future. In the meantime, however, a balanced exploration, acquisition and development program for cobalt is necessary to reduce the vulnerability of the U.S. to disruptions in its supplies of the metal.

24
Mercury and Cinnabar

MERCURY (Hg) IS ALSO CALLED "quicksilver," and it is one of the very few metals that is a liquid at room temperatures. In the metallic form, it has a silvery-white color with a faint bluish tinge. The metal is a white solid when it is below its melting point of -38.87 degrees Centigrade, and above its boiling point of 357 degrees Centigrade it is a colorless vapor. However, small quantities of mercury will volatize at room temperatures. Its other properties include:
- High density (specific gravity about 13.54 at room temperature).
- Uniform volume expansion in its liquid state with increasing temperatures.
- Good electrical conductivity.
- The ability to alloy (amalgamate) readily.
- A high surface tension; chemical stability.
- The toxicity of its vapor and most of its compounds.

Mercury occurs primarily in combination with sulfur to form more than 25 different minerals, but the most important commercial ore is the red mercuric sulfide mineral called cinnabar (HgS), which contains 86.2% mercury and 13.8% sulfur. The liquid native metal occurs in some ore deposits, and mercury has also been recovered from other ore minerals. Among these are corderoite ($Hg_3S_2Cl_2$), Livingstonite ($HGSb_4S_7$), calomel (HgCl), metacinnabar (a black or gray form of cinnabar) and other mercury minerals.

Marcasite, pyrite, stibnite, and small quantities of other sulfide minerals are often associated with mercury ores. Except for gold and silver, there is a notable lack of association between most cinnabar deposits and other primary ore minerals.

Most cinnabar ore deposits occur at fairly shallow depths, and the predominant number of economic deposits are located almost entirely in regions of late Tertiary orogeny that were formed during periods of intense volcanic activity. Cinnabar ores are also commonly found in areas where hot spring activity, past and present, is evident. Among the more common host rocks for cinnabar are limestone, calcareous shales, sandstones, serpentine, chert, andesite, basalt and rhyolite.

The predominant gangue minerals are opal, chalcedony and

quartz, sometimes calcite or dolomite, and more rarely, barite and alunite. The replacement of adjoining country rock formations by dolomitic carbonates or by opal is fairly common, but the cinnabar almost always occurs in these minerals (or in unaltered rocks) that are directly associated with quartz or chalcedonic silica. Hydrocarbons are also frequently present in the mineralized rocks.

Cinnabar ores may occur in irregular and chambered veins and brecciated zones, in stockworks made up of a multitude of minute seams, or as disseminated mineralization and replacements in the more porous rocks.

The predominant irregularity and brecciated character of most ore deposits indicates their formation and development near the surface of the ground. Most of the ore deposits are characterized by the presence of impervious "roofs" or "caps," which have been responsible for the localization of the ore shoots and zones of replacement or alteration.

Even though most known mercury deposits have been formed during relatively late geological time periods and primarily in association with the eruption of Tertiary and Recent lavas, it does not necessarily follow that ore deposition has been confined to the later geological times. Many of the older surface volcanic eruptions were undoubtedly accompanied in some regions by the formation of mercury ore deposits, which have since been eroded and dispersed.

Most mercury ore deposits have been formed at depths of less than one thousand feet, and very few individual deposits have been mined profitably at depths greater than 1,500 feet. However, a maximum depth of 2,400 feet has been mined in California, and the Almaden Mine in Spain is being mined to depths exceeding 2,430 feet.

Cinnabar ore deposits have been divided into two general types. The first type is disseminated ore bodies, in which the cinnabar has impregnated a more or less fine-grained or highly brecciated gangue rock. The second type consists of ores deposited in the fissures and cracks in unaltered country rocks as veins and bodies of almost pure cinnabar. The fairly open-textured sandstones and coarse breccias contain a multitude of open spaces that are conducive to the development of large, higher-grade ore bodies, whereas the finer-grained rocks such as shales or schists provide little room for deposition except along fissures or fractures in the rocks.

There are, however, other types of mercury ore deposits that are somewhat unique in certain areas or regions. For example, the ore body at the McDermitt Mine in Nevada was formed in ancient lake-bed sediments that are predominantly composed of volcanic ash which accumulated during the Miocene epoch. Most of the ore-grade mineralization is located within 165 feet of the ground surface, and the ore minerals consist of about 70% cinnabar and about 30% corderoite (mercury sulfochloride).

Since it is by far the most important ore mineral, cinnabar is described here as follows: a mercuric sulfide in the form $3(HgS)$, this mineral forms exagonal, rhombohedral, thick tabular or stout-to-slender prismatic crystals. It also occurs as crystalline encrustations, earthy coatings as disseminated granules, and in granular masses. Its color is cochineal-red, grading to brownish-red and lead-grey (metacinnabar). Cinnabar has an adamantine luster, which inclines from metallic when dark-colored to dull in friable varieties; it gives a scarlet streak and has a subconchoidal, uneven fracture. Its hardness is between 2.0 and 2.5 and its specific gravity is 8.09.

Lindgren notes that native quicksilver, calomel ($HgCl$), montroydite (HgO), and several of the oxychlorides are evidently secondary minerals which developed from the primary sulfide. Some other primary (but fairly rare) mercury minerals are: the black telluride, coloradoite; the selenide, teimannite; the sulfoselenide, onofrite; and other much rarer combinations of the selenides of copper, lead, and mercury. A mercury-containing tetrahedrite has also been identified.

Although cinnabar ore deposits generally form a well-defined group, they are not entirely separated from other classes of ore deposits. Some of the cinnabar deposits contain other metallic minerals and there are several that show a transition to stibnite and arsenical ores. In Nevada particularly, there are many instances of a close relationship between gold and silver ore bodies and cinnabar deposits, and there are also several associations with stibnite (antimony sulfide) deposits.

It has been noted that hot springs and volcanic surface flows are present in nearly all regions that contain economically important mercury deposits, excepting the major ore deposits at Idria and Almaden. Considerable quantities of cinnabar have been found to be closely associated with hot springs deposits, and the mineral has also been found where it is currently being deposited by hot springs. Lindgren says this indicates that hot springs solutions have formed the majority of cinnabar deposits and, for the few deposits where no such clear association is evident, the characteristic mineral association still holds good—which indicates that volcanic activity and hot springs actions cause the formation of these deposits as well. He also pointed out the fact that signs of igneous activity may not have reached the ground surface and that older hot springs activity may have subsided.

Prospecting for mercury ores involves the discovery of mineralized outcrops by visual means, the tracing of pieces of float rock, or by the panning of eroded materials or crushed rocks. Because of its high specific gravity, cinnabar (and most other mercury minerals) can be easily separated in the miner's gold pan. The mineral "tails" much like gold particles in the pan, and its distinctive reddish coloration provides for an easy means of identification.

As a result, most suspected cinnabar-bearing outcrops can be easily checked by panning eroded materials and soils located immediately below the outcrop. Alternatively, pieces of higher-grade mineralization can be crushed with a mortar and pestle, and the resulting powder panned by hand. If the mercury mineralization appears to be fairly widespread, eroded sand and gravels from washes, ravines and hillside runlets can be panned to see if they contain cinnabar particles.

Because of the fairly frequent association of mercury with gold and silver deposits, areas with identified mercury mineralization should also be checked for the potential occurrence of precious metals. Although geochemical methods have seldom been used for the primary purpose of identifying mercury deposits, they are commonly used to identify the presence of mercury as an indicator element for the possible occurrence of precious metals deposits.

The sometimes-close association of mercury with precious metals ores has resulted in the production of significant quantities of mercury as a byproduct of mining for gold and silver at some locations. For example, a large amount of byproduct mercury has been produced from these major gold mines in Nevada: Alligator Ridge in White Pine County, the Borealis Project in Mineral County, the Carlin

Mines complex in Eureka County, the Hog Ranch Mine in Washoe County, the Jerritt Canyon Mine in Elko County, the Paradise Peak Mine in Nye County, and the Pinson and Prebble mines in Humboldt County.

Other examples of mines that have produced significant amounts of byproduct mercury are the McLaughlin Mine in Napa County, California, and the Mercur Mine in Tooele County, Utah.

25
Tantalum and Columbium

TANTALUM AND COLUMBIUM are almost always found in nature as oxides in association with other minerals in complex molecular form, and not in elemental (native) form or as sulfides. The average crustal abundance of columbium is estimated at about 24 parts per million and the average abundance of tantalum is about 2.1 parts per million. Columbium is called "niobium" (symbol Nb) in the international community.

Tantalum and columbium have a strong geochemical coherence, are closely associated with each other, and are commonly found together in most rocks and minerals in which they occur. Both metals primarily occur in nature as oxides, multiple oxides and hydroxides. The principal ore minerals currently being mined are pyrochlore, pandaite, tantalite, columbite, loparite and ixiolite.

Columbium and tantalum minerals are known to occur in placer deposits at Bear Valley and Dismal Swamp in Idaho, where they have been mined in the past, but the tantalum content of these deposits is low. Some of the other known occurrences of columbium and tantalum minerals include deposits in pegmatite formations and placer gravels located in California, Arizona, Colorado, Maine, North Carolina, South Carolina, South Dakota, Utah, New Mexico and Alaska.

At the time of this writing (late 1991), the United States was dependent upon foreign sources for all of its supply of columbium and most of its supply of tantalum, and there had been no significant tantalum-columbium mining industry in the U.S. since 1959. Both metals are normally produced and sold in concentrates (60% minimum). However, most identified U.S. tantalum/columbium resources to date are fairly small and/or low-grade.

Considering the heavy U.S. import reliance for both metals, their average price, and increasing demand, the opportunities for prospecting and developing economical tantalum-columbium deposits in the U.S. are very favorable—especially for low-cost, small-scale mining operations.

The world's known economic concentrations of columbium and tantalum minerals occur primarily in the following related members of intrusive alkaline rock complexes: carbonatites, alkaline granites, pegmatites and nepheline syenites.

Pegmatites and their associated placer deposits once served as the major source for tantalum and columbium, which were contained primarily in the minerals tantalite and columbite. At present, pegmatites and their respective placer deposits contribute less than 5% of the market economy countries' supplies of columbium and tantalum, but this source of supply could become significant in the future.

Tantalite and columbite occur mainly as accessory minerals that are disseminated in granitic rocks or in pegmatites associated with granites. Economic mineral concentrations have occurred where weathering and deep erosion have led to the formation of residual or alluvial deposits, or where the pegmatites contain greater concentrations of the ore minerals as a result of natural crystallization processes.

The microlite-pyrochlore mineral series is also a source of tantalum and columbium. These minerals essentially consist of complex oxides of tantalum, columbium, sodium and calcium in combination with hydroxyl (OH) ions and fluorides. Unlike columbite or tantalite, most of these minerals are close to one end of the series or the other. Microlite occurs mainly in the albitized zones of granite pegmatites and is often associated with tantalite or columbite.

Both columbium and tantalum are often produced as byproduct commodities in the mining of tin deposits, such as the operations on the Jos Plateau in Nigeria. Struverite is a low-grade source of tantalum and columbium that is recovered from tin mining wastes in Southeast Asia. The struverite mineral is a variation of the titanium mineral called rutile, and it typically contains about 12% of each of the tantalum and columbium pentoxides. Prospecting for primary tantalum minerals is generally pursued on the basis of the metal's known frequent association with tin and certain other elements in pegmatite formations, some features of which can be distinguished by aerial photography. The high specific gravity of tantalum minerals makes it possible to detect their presence in placer deposits by panning the materials in a gold pan.

Columbium deposits of economic importance are most likely to be found in alkali rock complexes and the associated carbonatites. Alkalic rock complexes are usually restricted to the more stable Precambrian cratons of the continental assemblages. These complexes are often located in spatial relationships with fault lineaments, such as the Kapuskasing-Moosonee High and the St. Lawrence River Fault in Canada, and the marginal fault of the East African rift system. However, some occurrences are less obvious, particularly where the spatial relationships may not exist. Alkaline complexes commonly occur as small circular or elliptical bodies (usually less than 5 miles in diameter), arcuate ring-like structures, cone sheets, or crosscutting dikes.

Carbonatite rocks are the major source of the world's known columbium occurrences, and they currently account for almost all of total production. The carbonatites usually occur towards the center of alkaline complexes in the form of plugs or irregularly shaped bodies of up to 3 square miles in area, or as a crosscutting dike. The mineralization consists of chemically unique igneous rocks that contain anomalous quantities of carbonates, and they often resemble marble (calcium carbonate).

A majority of the carbonatites contain an excess quantity of calcite and are called sovites. Those that are dominated by magnesium carbonate content are called ankerites. The mineral pyrochlore characteristically reaches its highest concentration in sovites, although it has been found in almost all of the associated rock types of the alkaline granitic suite.

Phrochlore is a complex molecular combination of several elements as is its barium analog, pandaite, and both minerals have become the major source of columbium. Pyrochlore typically contains from 10% to 15% calcium oxides and very little barium oxide, whereas this proportion is nearly reversed in pandaite. Both of the minerals have a low tantalum content, with a columbium-to-tantalum-oxide ratio of 200 to 1 or greater.

Pyrochlore and pandaite are commonly found in the interior zones of alkali rock complexes, and they are frequently associated with minerals containing such other elements as titanium, thorium, uranium and the rare earth metals (lanthanide series) as well as phosphates and fluorite.

The combination of radioactive thorium and uranium content has aided in the discovery of carbonatite deposits. Radiometric reconnaissance, using a gamma ray scintillometer, was instrumental in discovering the Niobec ore body. In addition, magnetometer surveys have also proven useful in the discovery of carbonatites that have been enriched in magnetite.

Canadian pyrochlore deposits occur in complex ring structures of the carbonatite and alkali rocks in the Precambrian Shield. In Brazil, the occurrences are primarily in eluvial deposits that have resulted from the weathering in-situ of syenite-carbonatite rocks, which left an enriched concentration of magnetite, apatite and pyrochlore.

Nepheline syenites tend to be somewhat enriched in columbium and tantalum content, and they are especially prevalent in the [former] Soviet Union. The principal columbium mineral in the great Lovozero alkali massif of the Kola Peninsula is loparite, which is a cerium- and columbium-bearing variety of the titanium mineral, perovskite.

Pyrochlore mineralization can occur in either primary or secondary carbonatites. The primary deposits are represented by hardrock ore bodies that are normally found in the more temperate climates (such as the U.S.), where weathering has not been a significant factor in concentration. In this type of deposit the columbium occurs primarily in the mineral pyrochlore, in concentrations ranging from about 0.5% to 0.7%. Both the Oka and Niobec deposits in Canada are examples of this type.

Secondary deposits are usually found in tropical regions that are subjected to an abundant annual rainfall. Deep, in-situ weathering of the host rocks results in a strong residual enrichment of the more resistant columbium, phosphate and iron minerals. Under these conditions, the columbium may occur in an altered mineral form of pyrochlore called pandaite (much more appropriately called bariopyrochlore), in which the sodium and calcium ions have been replaced by barium and/or strontium in varying percentages.

In these deposits, enrichment of the columbium content is from about 2 to 10 times greater than the quantities found in primary ores. The highest concentration of columbium ore is found in the eluvial or laterized materials that form a cap overlying the carbonatite deposit. All of the known Brazilian deposits have undergone this columbium enrichment process, and African deposits, particularly those in Zaire, have undergone similar enrichment, but to a lesser extent.

The importance of this enrichment in grade can be illustrated by a comparison of the in-situ demonstrated ore resource with the amount of contained columbium. As an example, Brazil presently has about 22% of the world's known in-situ ore resources, but these deposits contain nearly 60% of the total known reserves of columbium.

Alkaline granites and related ultramafic rock formations are a potential source of both tantalum and columbium in Canada. The deposits are apparently associated with later stages of the alkaline intrusive process, and they typically contain a much higher concentration of tantalum in the columbite-tantalite mineralization. The Crevier and Thor Lake deposits are good examples of this type of occurrence.

26
Tellurium: Properties, Ores, and Assay Procedures

Author's Note: The following information is intended to provide a basic understanding of tellurium and its most notable compounds and occurrences. A general description of the element's properties and distribution will be followed by a delineation of tellurium's most important occurrences both as an ore-forming mineral and as a byproduct mineral. In addition, several of the tests and assay procedures used for tellurium and its compounds will be described in detail.

TELLURIUM (SYMBOL: Te) exhibits both metallic and nonmetallic chemical properties, is tin-white in color, has a metallic luster, and occasionally occurs in the native state and as a dioxide. The element has an atomic weight of 127.60 and an atomic number of 52, and it is chemically similar to selenium and sulfur. The pure element has a hardness of between 2.0 and 2.5, a specific gravity of 6.1 to 6.3, and a melting point of 451 degrees Centigrade.

Tellurium has been found primarily in combination with such metals as bismuth, cobalt, copper, gold, mercury, nickel, lead and silver. As a group, these compounds are known as the tellurides.

In the native state, tellurium forms hexagonal, prismatic or acicular crystals and it has also been found in columnar or granular masses. The native element has the same tin-white color and metallic luster, hardness and specific gravity as the refined product. Tellurium dioxide forms orthorhombic acicular crystals, and it has also been found in spherical masses. The dioxide has a white color, subadamantine luster, hardness of 2.0 and SG of 5.9. Both forms of the element have been found in California, Colorado, Nevada, Rumania and Western Australia. To date, neither form has been found in quantities large enough to provide a source for supplies of the element.

The few estimates available on the quantity of tellurium occurring in the earth's crust show it in the range of 0.5 to 10 parts per billion (ppb), with several tables showing the amount as .002 parts per million (ppm). The relative abundance of tellurium in U.S. porphyry copper ores has been estimated at 1,000 ppb (1 ppm). There are at present no known mineral deposits primarily minable for their tellurium content, and most of the recoverable metallic resource is believed to be contained in the world's copper resources. It occurs as tellurides associated with copper sulfides and copper-nickel sulfides in these deposits. The element has been recovered mainly from the precious metals-rich anode slimes that accumulate during the electrolytic refining of copper.

Potentially recoverable tellurium occurs in coal deposits, manganese nodules, and gold, silver and lead ores. Lead ores contain about one-fourth of the quantity available from copper ores while coal contains about four times as much as the copper ores.

Although the tellurides of gold and silver are ore minerals in some deposits of these metals, they have not been previously used as sources of the element. However, the development of the Emperor telluride leaching process in Fiji has made the economic extraction of tellurium from gold ores feasible. The process involves the selective flotation of the crushed ores to produce a tellurium concentrate, with the gold going off in the rougher tails. The concentrate is then chemically treated to produce a high-purity tellurium byproduct.

The presence of tellurium in complex ores, particularly those that contain the precious metals, is commonly believed to cause several unusual problems in the identification, assaying and processing of the economic minerals contained in these types of ores. Historically, some of these problems were based upon the actual difficulties encountered when work was first attempted on compounds containing the element. However, with the advances in technology over the last century it has been shown that many of the problems previously thought to be associated with the presence of tellurides were actually due to other causes. These include basic errors in procedures, incorrect sampling techniques, or the presence of other problem mineral components.

Certain losses can occasionally occur when assaying telluride ores, but these have been proven to be fairly small in relation to the metallic content of the samples and the errors can be easily corrected if the presence of tellurium or other contaminants is known. Several of the proper tests and procedures are outlined later in the article.

Common Telluride Ores

A basic familiarity with the physical nature of the significant telluride ores is an indispensable aid both in their identification as primary ore minerals and in the detection of their presence in complex ores and disseminated mineral deposits. A brief description of the most important ores, their distinguishing characteristics and their most notable occurrences is given below.

Calaverite is a gold ditelluride that sometimes contains silver. It forms monoclinic, bladed crystals and also occurs in granular masses. Brass-yellow to silver-white in color, it has a metallic luster, hardness 2.5 to 3.0, and SG 9.24. Calaverite been found in California, Colorado, Ontario (Canada), the Philippine Islands, and Western Australia.

Hessite, a silver telluride that forms cubic, massive crystals, is lead-gray in color. It has a metallic luster, hardness of 2.5, and SG 8.4. Hessite has been found in California, Ontario, Chile, Mexico and Australia.

Nagyagite is a lead-gold tellurium and antimony sulfide. It forms monoclinic, tabular crystals and also occurs in granular masses. Its color is blackish-gray and it has a metallic luster, hardness between 1.0 and 1.5 and SG 7.4. The mineral occurs in Colorado, North Carolina, Ontario and Western Australia.

Petzite, a silver-gold telluride, forms cubic crystals but normally occurs in granular masses. It is steel-gray to iron-black in color. Petzite has a metallic luster, hardness 2.5 to 3.0, and SG 8.7 to 9.02. It has been discovered in California, Colorado, Ontario and Western Australia.

Sylvanite, a silver-gold telluride that forms monoclinic, prismatic or tabular crystals, also occurs in columnar or granular masses. Its color is steel-gray to silver-white and it has a metallic luster; its hardness is 1.5 to 2.0 and SG is 8.16. Sylvanite occurs in California, Colorado, Ontario, Rumania and Western Australia.

Altaite is a lead telluride that usually occurs as granular masses and sometimes forms cubic crystals. Its color is tin-white with a yellowish tinge and it has a metallic luster, hardness 3.0 and SG 8.15. Usually occurring in association with native gold, other tellurides and sulfides, it is found in Colorado, North Carolina, British Columbia, Chile, Siberia and Western Australia.

Some of the lesser-known tellurides are: krennerite, a gold telluride similar to calaverite but occurring in orthorhombic crystals; tapalpite, a bismuth-silver telluride found in Mexico, and kalgoorlite, an iron-black mercury-gold-silver telluride found in Western Australia.

Simple Tests for Tellurium

A number of the standard reference manuals on assaying, mineralogy and metallurgy that have been published within the past 70 or so years are still being used by many professionals for performing simple tests to detect the presence or absence of tellurides in ore samples. Some of the most simple and proven methods used for this purpose are as follows:

The first, and easiest, method is to take some of the finely pulverized ore and heat it gently with a concentrated solution of sulfuric acid in a white container. If tellurium is present in the sample, a faint purple tinge will be seen surrounding the ore particles that will gradually spread through the solution, coloring it a deep carmine color if much tellurium is present. The color will disappear if the solution is boiled or water is added, both of which will precipitate the metal as a grayish-black powder. If the test produces doubtful results, take about 100 grams or more of the powder and pan it in a gold pan, and then treat the concentrate in the same manner.

When the blowpipe method is used for analysis, there are three simple tests that can be used on the sublimate:

1) Sublimate on a plaster tablet: Tellurides heated as-is, or with a bismuth flux on a plaster support, will yield a purplish-brown coating. A drop of concentrated sulfuric acid added to the film and gently heated will form a pink spot.

2) Sublimate on charcoal: When heated on charcoal, a white sublimate of tellurium oxide resembling antimony oxide is formed near the assay. The coating is volatile, and when touched with a reducing flame it will color the flame a pale green.

3) Test with sulfuric acid (as above): When gently warmed with concentrated sulfuric acid, powdered tellurides produce a reddish-violet solution. Too intense a heat or added water will cause the color to disappear.

Assaying Problems and Solutions

Some actual difficulties were encountered by assayers and chemists who first dealt with telluride ores. Lodge stated in 1910:

> Much gold can be lost when roasting ores rich in both gold and tellurium. Low-grade ores, when little tellurium is present, can be roasted with little loss of gold, and assays on these roasted ores will always run

much more uniformly than upon raw ores.

However, Edward E. Bugbee (1941) disagrees to some extent in his *A Textbook of Fire Assaying*. This particular book is a classic in its field and is still being used as a standard assay reference by the U.S. Bureau of Mines [abolished in 1996] and the U.S. Geological Survey.

In Chapter IX, "The Assay of Complex Ores and Special Methods," Bugbee notes that the assaying of ores containing tellurides of the precious metals had always been considered more than ordinarily difficult, and that some of the results obtained by different assayers (and even duplicate assays run by the same person) often varied considerably. However, he says that most of the differences reported were due more to difficulties arising from improper sampling than from any chemical interference due to the presence of tellurium.

Bugbee goes on to say that a telluride mineral may contain as much as 40 percent gold, so that one 100-mesh particle added to or subtracted from the assay could make a difference of several hundredths of an ounce per ton in the final results. To eliminate this possibility, he recommends that all telluride ores be ground to a fineness of around 200 mesh and be thoroughly mixed before assaying. Of course, this doesn't help much if the assayer does not suspect or test for the presence of tellurides.

The same text observes that the presence of tellurium in a lead button will reduce the surface tension of the button, allowing it to "wet" the surface of the cupel. This partial dispersion allows some of the material to be absorbed into the cupel while other portions of the alloy form minute beads that are left behind on its surface.

Bugbee states that if lead is added to the alloy in an amount about equal to 80 times the amount of tellurium present in the button, then the precious metals losses become virtually negligible. He said assayers should allow for large lead buttons to eliminate any problems associated with tellurium. He also recommends the use of a cupel that has a finer surface when cupeling buttons that contain tellurium in order to eliminate any possible absorption. In addition, the manual also mentions that silver in a button ally cuts gold losses that result from the presence of tellurium. Silver acts as a diluent for gold and, Bugbee states, should always be added to all gold assays for this reason if for no other.

Finally, Bugbee notes that most authorities feel the scorification process is not at all reliable for assaying telluride ores. Hillebrand and Allen (1905) devised an assay charge for ores containing from 15 to 19 ounces of gold and from .074 to .092 percent tellurium, in these quantities: Ore, one assay-ton; sodium carbonate, 30 grams; borax glass, 10 grams; litharge, 180 grams; reducing agent for 25-gram buttons; and silver, from 2.5 to 3 times the gold content. With this assay charge, slag losses are no higher than those encountered with ordinary gold ores, and there are no significant losses during cupellation. They also recommend that for ores that contain higher amounts of tellurium, then the quantity taken for assay should be reduced and the balance of the assay charge given above should remain the same.

27
The Geology and Technology of Thorium

THORIUM IS A GRAYISH-COLORED and radioactive metallic element of the actinide series, which only occurs in small quantities in certain rare minerals. It has an atomic weight of 232.05, atomic number 90 and chemical symbol Th, and it has chemical properties similar to titanium, zirconium, and hafnium. Slightly radioactive, with a half-life of 1.39×10^{10} years, this element decays through 12 intermediate radioactive isotopes to finally yield a stable isotope of lead.

Thorium is widely distributed in the earth's crust and is usually associated with uranium or the rare earth elements. Its geochemical abundance is estimated at about 5.8 grams per metric ton in the upper layers of the earth's crust, and it is about three times as plentiful as uranium. Monazite in beach sand placers is the principal source of the metal. It is also found in mineralized veins, deposits in sedimentary rocks (such as thorium-bearing dolomite), in conglomerates or quartzites that have undergone enrichment, and in certain uranium ores.

Thorianite (thorium oxide), thorite (thorium silicate), and monazite, a phosphate of the rare earth elements, are the principal thorium-bearing minerals. Thorianite and thorite are the only true thorium minerals, but monazite has been the primary source of thorium to date.

The mineral monazite may contain up to 8% thorium oxides, with the typical average thorium oxide content of monazite sands from various locations as follows: Malaysia and India, 8%; Malagasy Republic and Australia, between 7% and 8%; Brazil, Indonesia, Korea, Nigeria and South Africa, between 6% and 7%, and most monazite sands from the U.S. average, about 4%.

Thorium derived from monazite was produced in North and South Carolina as early as 1893. Monazite was also produced from placers in Idaho's Bear Valley during the 1950s. Titanium Enterprises processed monazite from tailings piles during 1979 and restarted dredging operations on beach sands in early 1980. The company recovered monazite as a byproduct while mining titanium and zirconium minerals from Pleistocene beach sands in northeastern Florida.

The thorium industry has historically been a small segment of mining operations that primarily recover other mineral commodities, such as ilmenite, rutile, monazite, and cassiterite, and in the production of other chemical compounds, such as the rare earth oxides. In Canada, thorium has been produced as a byproduct of uranium mining and processing; most of the world's thorium supply, however, has come from placer mining, and most of this production derives from the mining of beach sands.

Chemical separation of thorium is accomplished by digestion of the monazite with sodium hydroxide, which solubilizes the phosphate and leaves hydrous oxides of thorium, uranium and the rare earth elements after filtration. This residue is dissolved in hydrochloric acid, and the resulting solution neutralized with sodium hydroxide. All of the thorium and uranium salts are precipitated and contain only a small quantity of the rare earth elements. The crude thorium hydroxide is then dissolved in nitric acid and purified by solvent extraction.

In a modification of the acid process, thorium and the rare earth elements are precipitated from a diluted sulfate solution with sodium oxalate, leaving uranium to be recovered with an anion-exchange resin. The oxalate precipitate is treated with sodium hydroxide, yielding a hydroxide precipitate. The hydroxide is calcined to oxides, which are then dissolved in nitric acid for purification of thorium by solvent extraction.

Thorium and yttrium are also recoverable from thorite ores by leaching with sulfuric acid. The thorium is removed from the leach solution by solvent extraction and then stripped from the organic solvent by using ammonium carbonate solution. Thorium oxide is prepared by calcining the thorium oxalate precipitated from the nitrate solution that is derived from the solvent extraction process.

Production of high-purity thorium metal is difficult because it is very reactive and has a high melting point of 1,840 degrees Centigrade. The metal is prepared by reduction of the halides with calcium of magnesium, reduction of thorium oxide with calcium or sodium, fused-salt electrolysis, or thermal decomposition of thorium tetrachloride. Thorium prices are usually based on dollars per pound of thorium contained in the oxides.

Thorium compounds, such as the oxide, nitrate, and chloride, are the principal finished products. The oxide, thoria, is the most stable of the refractory oxides and is also used to a minor degree as a catalyst in the chemical and petroleum industries. Thorium compounds are used in the fabrication of incandescent gas mantles, at the rate of about one pound per 1,000 mantles.

Magnesium-based alloys containing varying percentages of thorium are used for aircraft and aerospace applications, and small quantities of the metal are used in producing certain metal alloys that are high-strength and corrosion-resistant. Metallic thorium and the oxide are used in computers, radiation detection devices, and electrical discharge tubes. The nitrate is also used in the manufacture of tungsten welding rods, and small quantities of thorium metal are used for nuclear fuels.

Operating factors in the thorium industry relate largely to the safety and environmental aspects. Environmental concerns must be considered in the mining of placer material, though this is not significant in the mining of beach sands since only an average of about 2% of the material is removed as concentrate, while the remainder is redeposited.

Special problems arise in the handling of thorium products because of their radioactivity, being similar to those involving the handling of uranium and other radioactive materials.

According to the [former] U.S. Bureau of Mines, there are many large deposits of thorium-bearing materials occurring in the United States, mainly in beach and stream placers, mineralized veins and carbonatites. It is also reported that there are no satisfactory alternative materials that are suitable for the major non-energy uses for thorium.

28
The Titanium Minerals

TITANIUM IS THE NINTH-MOST ABUNDANT element, and it comprises roughly 0.6 percent of the earth's crust. The metal occurs in nature only in chemical combination, most commonly with iron and oxygen. Even though titanium is relatively abundant in comparison with several other elements, economic ore bodies and placer deposits are not that easy to find. As a result, the average price paid for sponge titanium metal over the past few years has been about $5 per pound.

Since the element is a fairly common constituent of beach sands and older alluvial placers, as well as a few large ore deposits, its commercial value makes the metal worthy of additional attention for the serious prospector. In addition, the United States has relied on foreign imports for about three-quarters of its consumption of titanium dioxide in concentrates in recent years. Because of this import reliance, the FY1989 Defense Appropriations Bill (House bill 4781-13) appropriated $6 million for the Title III Program of the Defense Production Act to "develop a reliable supply of titanium ore from ilmenite." A major objective of this appropriation is to help reduce U.S. import reliance for titanium concentrates, which was about 80% of U.S. consumption in 1988.

The most common titanium mineral is ilmenite, which is a titanium-iron oxide that has an average titanium dioxide content of over 50%. The name is also used to describe placer materials that have been oxidized and leached by weathering, which can contain up to 70% titanium dioxide. Extreme alteration of ilmenite produces an amorphous-to-finely-crystalline TiO_2 called leucoxene.

Ilmenite occurs in hexagonal, tabular crystals and in masses, and it has a black color and metallic luster. The mineral has a conchoidal-to-subconchoidal fracture with no cleavage, a hardness between 5.0 and 6.0, and a specific gravity of about 4.75. It occurs as an accessory mineral in gabbros, diorite, anorthosite and pegmatite, as well as certain ore veins and many placer deposits.

Predominantly a crystalline titanium dioxide, rutile is another important titanium mineral. Synthetic rutile, or rutile substitutes, are often derived from ilmenite by processes utilizing oxidation-reduction treatments and the leaching of the iron content.

Rutile occurs in prismatic or acicular crystals and also in granular masses. Its color varies from yellow through reddish-brown to blue or black, with a metallic adamantine luster. The mineral has a conchoidal to uneven fracture with good cleavage in one direction, a hardness of 6.0 to 6.5, and a specific gravity of about 4.23. Rutile occurs in gneisses, schists, plutonic igneous rocks, and pegmatites.

Anatase is also a crystalline titanium dioxide, but it has a different crystal structure than rutile. The mineral has not been produced commercially in the past, but large deposits of anatase-bearing ores in Brazil are now being developed.

Anatase (also called dauphinite, hydrotitanite, octahedrite, oisanite, wiserine or xanthitane) occurs in acute pyramidic crystals that are often highly modified, and also in tabular crystals. The color is normally various shades of brown, grading to indigo-blue or black, but it is sometimes greenish, blue-green, pale lilac or slate grey. The mineral has an adamantine or metallic luster and a subconchoidal fracture, with a perfect cleavage in one direction. Its hardness varies from 5.0 to 6.5 and it has an average specific gravity of 3.90. Anatase usually occurs in vein or crevice deposits in gneiss and schists.

Nearly all of the commercially produced ilmenite and rutile came from older beach sands and river placers until the early 1940s, but by 1983 nearly 40% of the world's ilmenite was being produced from ore deposits. However, rutile is still being produced almost exclusively from beach sand deposits.

Most beach and stream placers commonly contain both rutile and ilmenite, with the titanium minerals being closely associated with magnetite, monazite, zircon, garnet, staurolite, sillimanite, chromite, gold and occasionally certain radioactive minerals. Because of their relatively high resistance to chemical decomposition and their hardness (5.0 to 6.5), ilmenite and rutile granules can survive the geological processes of disintegration, comminution, and the sorting action provided by moving water. As a result, they can eventually become concentrated in economic placer deposits. Ilmenite exhibits various degrees of alteration in sand and gravel deposits because of oxidation and the preferential leaching of iron content, so it may therefore vary widely in its average titanium dioxide content.

The Titanium Minerals

Almost all of the known commercially viable ore deposits of the titanium minerals are associated with anorthositic or gabbroic rocks, and they are usually of three main types: ilmenite-magnetite (titaniferous magnetite), ilmenite-hematite, and ilmenite-rutile. Ilmenite-magnetite deposits normally contain the ilmenite and magnetite as granular intergrowths that can be easily separated from each other, but they are sometimes found to be intergrown on a molecular level. Titanium minerals also occur in some carbonatite deposits, usually in the form of anatase or perovskite.

Sand deposits of titanium minerals are mostly located near continental margins, where erosion of regional granitic and metamorphic rocks containing ilmenite and rutile leads to the accumulation of these minerals in coastal plain sediments. The working and reworking of these sediments by ocean wave action on beaches and river deltas has resulted in various degrees of sorting and concentration by particle size, density and resistance to abrasion. Well-sorted sands, particularly in older deposits, are much more likely to contain concentrations of ilmenite, rutile and other heavy minerals than the more recent, relatively unsorted deposits.

Most titanium minerals are fairly dark in color, and their concentration in predominantly quartz sands is often very visible. Many of the beach sand placer deposits have been discovered through surface observation of higher-grade placer zones on beaches or along stream and river beds. These visible signs have sometimes been traced into some larger, lower-grade placer concentrations that have constituted economic ore bodies.

When prospecting for concentrations of heavier minerals, hand panning of samples has proven to be an excellent means to quickly produce a clean concentrate for examination. Ilmenite, rutile, gold and several other potentially valuable minerals can generally be identified by the use of a magnet and magnifying lens. Radioactive minerals, such as thorium-bearing monazite, can usually be identified by use of a Geiger counter or scintillometer. All samples used for panning and examination should be taken from the maximum depth possible in order to avoid misleading results caused by local surface variations in mineral content.

According to the [now defunct] U.S. Bureau of Mines, an approximate minimum requirement for an economic deposit of titanium minerals in the southeastern U.S. should consist of reserves of one million tons of titanium dioxide, an average grade of raw materials approaching 1% titanium dioxide, an average of 3% to 4% content of other heavy minerals, and an average depth of about 15 feet of economic placer material.

In 1989, ilmenite concentrates from beach sands were produced from three sites in Florida, and mineral sand deposits were being tested for ilmenite content in Dinwiddie and Sussex counties in Virginia. In addition, the U.S. Geological Survey has been investigating the offshore potential for titanium and zircon along the coastal areas of Virginia and South Carolina.

Large dredges are normally used to mine beach sand deposits containing titanium and other valuable minerals. Most of these are of the suction type, but a few of the bucket-ladder type dredges are still in use. The dredges are normally floated in their own pond, digging forward and stacking tailings behind. The de-sliming, removal of organic matter, and rough concentration of heavy minerals takes place on the dredge itself or on barges that are towed behind the unit. In a few operations where dredges are not practical, draglines are used to excavate the material and concentrates are produced in a permanently situated processing facility.

The rough concentration process, or the separation of heavier minerals from the lighter quartz-feldspar-mica fraction, is usually accomplished by wet-gravity methods. The Humphreys spiral has been the standard equipment used for concentration for over 40 years, but some pinched-sluice separators of various designs have also come into use. The Reichert cone concentrator, which operates on the pinched-sluice principle, has been used successfully for large-tonnage mining operations in Australia, and flotation has also been utilized under certain circumstances.

Jigs are normally used for separating heavier minerals from stream and river placer deposits because they are the concentrating device that is least sensitive to extreme variations in grain size.

The final wet-gravity concentrate is usually dried in a rotary kiln prior to further treatment, and the subsequent flow sheet depends almost entirely on the assemblage of minerals to be treated. The ilmenite and rutile are normally removed together by electrostatic separation, and this fraction is then subjected to a high-intensity magnetic separation to yield a final ilmenite product. The rutile fraction is further cleaned by screening and additional electrostatic separation. Zircon and monazite are recovered from the nonconductor fraction of the wet-gravity concentrate by a combination of gravity, electrostatic and high-intensity magnetic separation.

Most of the ilmenite from hardrock deposits is crushed and ground before being subjected to wet-gravity concentration. Magnetite is sometimes recovered by wet-magnetic separation, and an ilmenite concentrate is usually produced by flotation processes or wet-gravity methods.

The feed materials required for titanium sponge metal production are the same as for producing chloride-process pigments, because formation of titanium tetrachloride is required in both cases. Rutile is the only titanium raw material used for metal production, although ilmenite and titanium slags can also be used for this purpose.

After the metal has been converted from the oxide to the tetrachloride form, it is reduced with sodium or magnesion under an inert atmosphere at temperatures up to 1,040 degrees Centigrade. The residual chlorides are removed by vacuum distillation, inert gas sweep, or after cooling and crushing, by leaching in very dilute acid.

29
Tungsten Deposits

MOST OF THE WORLD'S TUNGSTEN production has come from ore bodies closely associated with contact metamorphic deposits called tactites (or skarns), hydrothermal vein deposits, stockworks and related deposits. Some other significant ore deposits have been found in pegmatites, associated with mineral hot springs, in placer deposits, and in certain lake brines. The temperature at which tungsten deposition occurs in different types of mineral deposits can vary considerably, so there is a good potential for the occurrence of several other metallic elements in tungsten deposits. In some cases, the value of the associated minerals can equal or exceed the value of the contained tungsten, thereby increasing the profit potential.

The primary source of tungsten ores in the United States is tactite (skarn) deposits, which have accounted for about 70% of the known domestic reserves of the metal. Tactites are garnet-bearing metamorphic rocks that have been formed by a high-temperature replacement and recrystallization of limestone or dolomite, and these deposits are emplaced at or near the contact with intrusive igneous rocks—usually granodiorite or diorite. Scheelite is the main tungsten mineral occurring in skarns or tactites, and it is not normally distributed evenly in the deposit. In most ore deposits, the scheelite occurs in fairly well-defined veins or shoots or is concentrated along bands in the host rocks.

The most important economic tungsten ores are scheelite (calcium tungstate), wolframite (iron-manganese tungstate), ferberite (iron tungstate), and huebnerite (manganese tungstate), in which the metal occurs principally as tungsten trioxide (WO_3). Tungsten is often found closely associated with varying amounts of copper, antimony, bismuth, tin or molybdenum. In some deposits, it is produced as a byproduct or coproduct in the mining of other mineral commodities.

Hydrothermal solutions are an essential ingredient in the formation of tungsten ore deposits. The metal gradually becomes concentrated in the residual fluids of crystallizing intrusive magmas as the tungstate ion, tungstic acid or sodium tungstate, and these chemicals are then precipitated in contact zones with the wall rocks, usually as scheelite or wolframite. The specific mineral formed is largely controlled by the relative amounts and activity of calcium, iron and manganese present in the molten solution. Although more than 20 separate tungsten-bearing minerals are presently known to occur, only the scheelite and wolframite groups have been of economic importance.

Scheelite is a heavy, white mineral with a nonmetallic luster, and it fluoresces from blue to yellow to white under an ultraviolet lamp — the color depending upon the impurities present. The mineral has a hardness of 4.5 to 5.0 and a specific gravity of 6.1. Because of its fairly high specific gravity, rocks containing even moderate quantities of scheelite are noticeably heavier than other rocks of comparable size. Scheelite is the main ore mineral in most tungsten deposits that have been found in California, Nevada, Montana and several other western states. Ore deposits of the mineral range in size from fairly small, isolated pods of ore spread along a contact zone to massive ore bodies in replacement deposits.

The wolframite mineral group comprises a solid solution series in which iron and manganese are the primary constituents of the tungsten compounds. The wolframite minerals are called "black ores" by the industry, and they are the predominant tungsten minerals found in quartz vein ore deposits. Wolframite has a color range from dark grey to brownish-black to black, a hardness of 4.0 to 4.5, and a specific gravity of 7.4. Ferberite and huebnerite have a similar hardness and specific gravity, so fairly small amounts present in the host rocks make them noticeably heavier. The minerals are often associated with the tin ore, cassiterite, as well as several other base metal compounds.

Tungsten-bearing quartz veins account for more than 60% of the known world reserves of the metal. Most of the quartz vein deposits are associated with intrusive igneous rocks of granitic composition, and the veins are usually located in the immediate vicinity of igneous intrusive contact zones. Quartz veins may contain both scheelite and wolframite with minor amounts of other contained minerals, but wolframite usually predominates in these types of deposits.

In contact-metamorphic tactite or skarn deposits, the scheelite is the predominant mineral, and it is usually associated with garnet, calcite, hornblende, pyroxene, diopside or epidote. Massive quantities of garnet are common, and the scheelite has a tendency to form along the periphery of limestone inclusions in the intrusive granodiorite as well as in strata in the intruded limestone that are favorable for deposition. Ore bodies also usually form along the contact zone as veins or bands in the host rock.

One example of an intrusive stockwork deposit occurs in the porphyry molybdenum ore body being mined at Climax, Colorado, where tungsten has been produced as a byproduct of mining operations. There were three separate intrusions in the formation of the moly deposit that created overlapping ore zones. Molybdenum-rich ore zones are overlain by zones that are poor in moly content but which contain significant quantities of wolframite. This type of association of tungsten with molybdenum in ore deposits of a similar nature has been found in several other localities.

Tungsten is processed to form a concentrate, and it is then sold in "units" of tungsten trioxide (WO_3) per ton of concentrate. A short ton unit of WO_3 is equal to 1% of a ton or 20 pounds of WO_3, which contains 15.86 pounds of tungsten metal. Internationally, tungsten concentrate is sold in metric ton units, 1% of which contains 7.93 kilograms of tungsten. In 1989, the average price paid per metric ton unit of WO_3 concentrate was $56; a 70% WO_3 concentrate would have sold for $3,920 per metric ton in the U.S. market. European market prices averaged $5 per unit higher, or $61 per metric ton unit of WO_3.

Some tungsten ores may be concentrated by using only a gravity separation circuit. Jigs and tables are especially useful in concentrating ores where the tungsten minerals occur in a coarse form, which permits the recovery of a coarse high-grade concentrate without incurring slime losses caused by overgrinding. If significant quantities of sulfide minerals are present in the ore, then a combination of flotation and magnetic separation procedures are used in conjunction with gravity separation. If valuable byproduct or coproduct minerals occur in the ores, they can be separated at some stage in the concentrating process. Most of the tungsten concentrates produced range from 50% to 70% tungsten trioxide content.

The WO_3 gravity concentration process and/or flotation process is hampered when excessive fines or slimes are produced by overgrinding; consequently, a primary consideration in ore processing is how to recover the tungsten with minimal crushing and grinding. This is accomplished by setting up the crushing and grinding in stages and to size the material at each stage, recovering as much of the tungsten minerals as possible from each size fraction. The normal mill circuit practice is to use screens, hydraulic classifiers, settlers and cones to obtain the highest possible recovery and produce the highest-grade concentrate at the same time.

30
Zirconium and Hafnium

ZIRCONIUM AND HAFNIUM are two chemically similar elements that usually occur together in several minerals. Rocks in the earth's crust have been estimated to contain about fifty times as much zirconium as hafnium, or a ratio of 50:1. Two minerals, zircon (zirconium silicate) and baddeleyite (zirconium oxide), are the primary source of both metals, and zircon is by far the most abundant. Zirconium- and hafnium-bearing minerals are rarely found in sufficient concentrations in rock formations to constitute minable ore deposits and are, therefore, primarily recovered from placer deposits.

Through gradual disintegration and erosion of the host rocks, the zirconium minerals are eventually concentrated by natural gravity processes into minable deposits of heavy, hard and chemically resistant sands. These sands usually contain appreciable concentrations of several other heavy minerals, including those that contain titanium and the rare earth elements. Placers of this type have been found in sand dunes, river bars and gravels and in ocean beach sands, the latter providing the major source of zircon, ilmenite, rutile and monazite.

Most of the beach sand deposits are mined by floating-suction or bucket-line dredges, the largest of which are capable of handling up to 1,400 tons of sand and gravel per hour. The finer material is processed by wet-gravity methods such as spiral concentrators, cones, jigs or sluices. These procedures produce a mixed concentrate of heavy minerals containing zircon, ilmenite and rutile (titanium-bearing minerals) and several other marketable minerals, such as monazite (a phosphate of the rare earth elements, yttrium and thorium).

The specific gravity of zircon (4.7) allows it to be easily concentrated with the other heavy minerals by gravity methods. Clean zircon sand is extracted from the heavy concentrate by a combination of drying, screening and electrostatic, electromagnetic and gravity separation techniques.

In contrast to ilmenite, rutile and many other heavy minerals, zircon is nonconductive and can be separated, together with monazite, by electrostatic methods. Monazite is slightly magnetic, and can be separated from the zircon sand by use of electromagnets. To obtain clean zircon, the zircon-rich concentrates from the electrostatic and electromagnetic processes are again subjected to gravity concentration with spirals or tables to eliminate the lighter and nonconductive minerals, such as quartz and kyanite. The enriched product is then usually dried at about 650 degrees Centigrade to remove organic materials, followed by treatment in high-tension separators and induced roll magnetic separators to remove any residual conductive and magnetic minerals. The final zircon product obtained by these procedures is usually 99% pure.

Approximately 95% of all zirconium consumption is in the form of zircon, zirconium oxide, or other zirconium compounds. The remainder is used as zirconium metal and in zirconium-containing alloys. The mineral zircon is primarily used for facings on foundry molds which, on the basis of actual zirconium content, account for more than 50% of the uses for the element.

Zircon facings on foundry molds increase their resistance to penetration by molten metals, and afford a uniform finish to metal castings. Milled or ground zircon is also used in refractory paints for coating the surface of casting molds. In the form of refractory bricks and blocks, zircon is widely used in furnaces and in hearths for containing molten metals. There are also many small but important applications for zirconium compounds in various chemical products.

Because of its low thermal neutron absorption cross-section and low radioactivity after radiation exposure, hafnium-free zirconium is used as cladding for nuclear fuel and as a structural material for nuclear reactors employing pressurized water heat exchangers. Both high-purity zirconium and zirconium-based alloys are used for these applications.

Small quantities of commercial-grade zirconium metal and alloys are used in the chemical industry as components in corrosive environments in heat exchangers, acid concentrators, tank shafts, valves, pump housings, fan wheels, high-speed agitators, electrode assemblies, steam jet exhausts, tubing, pipes and pipe fittings, spinnerets, and crucibles.

More than 85% of all the hafnium consumed is in the metallic form, most of which has been used to control rods in nuclear reactors for the U.S. Naval Reactor Program. A number of high-strength, oxidation-resistant alloys that contain appreciable quantities of hafnium are also used in gas turbine and jet engines, gun barrels, thermionic converters, reentry vehicles, and chemical processing equipment. The fastest-growing new application for hafnium is in the use of hafnium-columbium carbide for cutting-tool alloys, which provides a cost savings of over 30% or more when used instead of tantalum carbide. The remainder of the hafnium consumption is used primarily as the oxide in research applications.

Hafnium metal is used as control rods in reactors because of its good ductility and machinability, and its high thermal neutron absorption cross section. The metal also has excellent hot-water corrosion resistance. Because of these characteristics, hafnium is considered a preferred control rod material where relatively maintenance-free operation is necessary, such as in naval reactor applications. In addition, the absorption cross section of the metal does not decrease markedly after long periods of irradiation, because the absorption of a neutron by a hafnium atom produces several successive isotopes that also have large absorption cross sections.

Small additions of hafnium and carbon are made to tantalum, tungsten, columbium and molybdenum alloys to form a dispersed second phase that results in stronger high-temperature alloys. These alloys are used mainly in high-performance aircraft and in space power systems.

Australia leads the world in production of zircon, almost all of which comes from beach sands along the eastern coast (32%) and in Western Australia (68%). Production of zircon on the east coast, where it is a coproduct of rutile mining, has been declining because of lower grades and reserves, coupled with persistent environmental problems associated with mining beach sands. This loss has been largely compensated for by the increased production from Western Australia. South Africa normally ranks second as a producer of zircon.

Much of the world's zircon concentrate is a byproduct of mining operations for titanium minerals, excepting Malaysia and Thailand where it is a byproduct of tin mining. Several other countries that produce zirconium are the United States, Brazil, India, China, Sri Lanka, and Russia.

Most of the current U.S. domestic consumption of zirconium is provided by imports.

31
The Rare Earth Metals and Minerals

COMMONLY KNOWN AS the lanthanide series, the rare earth metals comprise a group of 15 chemically similar elements. These elements include lanthanum, cerium, praseodymium, neodymium, promethium, samarium, europium, gadolinium, terbium, dyspropsium, holmium, erbium, thulium, ytterbium and lutetium. Although it is not in the same group, yttrium is included as a rare earth element since it commonly occurs in the same minerals and also has similar chemical properties. Some of the elements in the rare earth group of metals are relatively abundant in the earth's crust, but minable concentrations are fairly uncommon.

Based on property differences due largely to varying ionic radii, the rare earth elements are classified into two groups: (1) the light (or cerium) subgroup, which comprises the first seven members (atomic numbers 57 through 63); and (2) the heavy (or yttrium) subgroup, which contains the remaining elements (atomic numbers 64 through 71) and yttrium (atomic number 39). In spite of its lower atomic weight, yttrium is included with the heavier elements because its ionic radius, occurrence and properties are much closer to the elements in that subgroup.

Industrial applications and demand for the rare earth metals and their compounds have undergone a dramatic increase as a result of the new technological advances made over the past 30 or so years. Fairly extensive research programs are continuing to create new applications and markets for these metals and their compounds, providing further stimulus to both production and demand. Since rare earth-bearing minerals occur in both lode and placer deposits, more miners are finding it worthwhile to become familiar with the various ore minerals and their identification.

The rare earth metals and yttrium are essential constituents in more than 100 minerals, but only a few of these minerals occur in concentrations that are sufficient to warrant their use as ores. At present, the minerals bastnasite and monazite are the primary mineral sources for these elements. Xenotime, an yttrium phosphate, is found in the same mineral environment as monazite, and it is currently the major source for yttrium. Apatite and multiple-oxide minerals such as euxenite and loparite are also commercial sources of these metals. A brief description of the most important rare earth-bearing minerals follows.

Monazite, a thorium-rare earth phosphate, forms prismatic crystals with a yellow, white or reddish-brown to brown color. It has a waxy or resinous luster, conchoidal to uneven fracture, hardness 5.0–5.5 and SG of 4.6–5.4. Monazite is found worldwide, primarily in placer deposits in quantities sufficient to constitute economic deposits.

Bastnasite is a fluocarbonate of the cerium subgroup of rare earth elements. At Mountain Pass, California, the mineral is associated with barium and strontium minerals, with the typical ore containing about 40% calcite, 25% barite and/or celestite, 10% strontianite, 12% bastnasite, 8% silica, and minor amounts of other minerals. The bastnasite mineral itself is composed of a mixture of about 75% rare earth oxides, primarily of lanthanum, cerium and neodymium (total about 94.5%), with minor amounts of the other rare earth elements. Bastnasite is found in lode deposits in carbonatites.

Xenotime. An yttrium phosphate that contains minor amounts of other rare earth metals, xenotime forms prismatic crystals with red, yellow or brown color and has a vitreous luster. It has uneven to splintery fracture, hardness 4.0–5.0, and SG 4.4–5.1. Xenotime been found in Colorado, Norway, Brazil and Malaysia.

Euxenite is a hydrated columbate and titanate-tantalate of uranium and several of the rare earth elements; it usually contains some thorium or calcium. Euxenite occurs in radial and parallel aggregates of prismatic crystals and also massive. It has a black, sometimes tinted, color and submetallic luster, subconchoidal to conchoidal fracture, hardness 5.5–6.5, and SG 4.7–5.0. Euxenite is found in Pennsylvania, Canada, Australia, South Africa, Finland, Norway, Sweden, Brazil and elsewhere.

Monazite can theoretically contain about 70% combined rare earth oxides (REO), which includes about 2% yttrium oxide, but most monazite concentrates contain from 55% to 65% REO. Processing companies usually specify a minimum of at least 55% contained REO. Bastnasite theoretically contains about 75% REO and very minor amounts of yttrium. However, most flotation concentrates average about 60% REO, which is upgraded to 70% REO by acid leaching and to 85% REO by a combination of leaching and calcining. Xenotime can theoretically contain about 67% yttrium oxides, but most of the concentrates average about 25%. The concentrates are usually upgraded to contain about 60% yttrium oxides and about 40% other REO.

When cerium (the most abundant of the group) has been separated, the remaining mixture of REO is called "didymium," and metal alloys made from such mixtures are called didymium metal or cerium-free mischmetal. The term "mischmetal" is used to describe a mixture of the rare earth metals in a metallic form.

Just two companies recovered rare earth ore minerals in the U.S. during 1987. Bastnasite was mined in California as a principal product and accounted for most of the U.S. rare earth production. Monazite was mined in Florida by placer methods, and it was primarily a byproduct of mining heavy mineral sands for titanium and zirconium minerals. Both thorium and yttrium were obtained as byproducts of the separation of rare earth metals from monazite.

Private foundations, educational institutions, research institutions, industry, several federal agencies and a number of foreign governments are presently conducting research on rare earth and yttrium applications. In addition, research on processing technology by the [since eliminated] U.S. Bureau of Mines involved improved beneficiation of bastnasite, greater energy efficiency and the recovery of byproduct minerals. Researchers in China are working to improve bastnasite recovery by developing a new flotation process.

Several private and government laboratories have been studying advanced rare earth separation techniques that could have commercial importance. Some processes that have been developed include: use of an ion-exchange chromatographic centrifuge, selective ionization of the elements with a laser, electromolecular propulsion, and altering oxidation states with ultraviolet laser energy. In the field, the U.S. Geological Survey is studying potentially significant concentrations of heavy mineral sands, including monazite, in surficial Atlantic Continental shelf sediments.

Except for the exploration for monazite deposits in the early years of the rare earth and thorium industries, most economic deposits have been discovered during the search for other minerals. Because some of the rare earth and yttrium minerals are naturally radioactive or are associated with radioactive elements, such as thorium and uranium, many of the discoveries have been made while prospecting for uranium. However, few of these deposits were economically feasible to mine at the time.

Most of the monazite deposits were discovered as a result of prospecting for gold, titanium, tin and zirconium in both alluvial and beach sand deposits. The large bastnasite deposit at Mountain Pass, California, was found by a prospector who was attracted to the area by the radioactivity of some minerals associated with the bastnasite. Current exploration techniques used for locating rare earth and yttrium minerals include the geologic identification of favorable mineral environments for deposition of these elements, as well as surface and airborne reconnaissance with radiometric and magnetometric devices.

Dredges, shaker tables, jigs and spiral concentrators are used to recover the heavier minerals, including monazite, xenotime, euxenite and other rare earth minerals, from placer deposits. The bulk of the heavy concentrate usually consists primarily of titanium and zirconium minerals, along with from 1 to 20% contained monazite. The monazite can be separated from the other minerals by a combination of electromagnetic, gravity and electrostatic recovery processes.

Froth flotation has been used to process Indian monazite and also for the vein monazite ores from South Africa. One flotation technique utilizes sodium silicate as a conditioner and sodium or potassium permanganate to improve the floatability of the monazite particles.

The bastnasite in California and China is mined by open-pit methods, while other deposits in New Mexico and Burundi have utilized underground mining methods. After the bastnasite ore has been crushed it is screened, hot-conditioned to improve flotation, and then fed to the flotation circuit. The concentrate from the flotation process is then thickened, drum-filtered and dried in a rotary kiln. This produces an average 60% REO concentrate that is upgraded to about 70% REO by leaching with 10% hydrochloric acid to remove calcite, and further upgraded to 85% REO by roasting the leached concentrate to remove carbon dioxide from the carbonate minerals.

The demand pattern for the rare earth metals and compounds is expected to continue a shift toward individual compounds, metals and special mixtures such as those used in X-ray screens, permanent magnets, fluorescent lamps, and electronic applications. It is also expected that improved knowledge of the properties of these elements will lead to several new industrial applications. The important industrial demand for individual metals having a high purity accounts for a large portion of the total value of each metal or compound because of the higher unit value, even though the amounts used are fairly small. The rare earth elements are currently used in metallurgy, electronics, nuclear devices, the chemical industry (mainly as catalysts), lasers, glass and ceramics, illumination, jewelry, and several other applications.

The higher relative value of some of the rare earth metals cannot be ignored, especially when they might be recovered as byproduct minerals in the concentrates produced in existing mining operations for other minerals. Despite the abundance of these elements, with the increased usage and growing demand they should prove to have a good future and a stable market.

Afterword

As a writer, researcher, consultant and mining professional, Dave W. Parkhurst would likely have been impressed by and appreciative of the tremendous volume and variety of information that today's Internet makes almost instantly available from one's computer, smartphone, or other modern communications device. The World Wide Web and the Internet were still in their infancy when Dave was putting in many hours driving to and from the post office, the library, county recorders' offices and other county offices, the BLM and various other federal agencies, state government and legislative offices, and other sources of information and assistance that he drew upon for his research and writing.

Helping to set up microwave towers was one of the jobs Dave had performed in the telecommunications industry prior to his becoming involved in the mining field full-time. He knew quite a bit about the emerging technologies in electronic communications, and perhaps he could envision to some extent how these technologies would revolutionize communication and the conduct of business—including the mining business—in the world by the advent of the twenty-first century and beyond. At the time of Dave's death, the use of Internet and cell phone communications was on the brink of becoming commonplace around the globe.

Other, mining-related technological developments that were still relatively new at the end of Dave's life included remote sensing, used to detect and map mineral formations below the earth, as well as related technologies that have facilitated the discovery and mapping of minerals in the U.S. and elsewhere.

Had he lived, Dave would have likely continued in the mining field in one capacity or another, ideally in minerals exploration. He would also have continued his lifelong pursuit of knowledge in his chosen field and in other areas of interest—something he enjoyed perhaps as much as being out in the hills "cracking rocks."

About the Articles' Author

David Walter (Dave W.) Parkhurst had several occupations in his lifetime, but it was metals and minerals exploration and mining that were his passions. From the time he was a child growing up in northern Nevada desert country to the last days of his life, he relished being in the outdoors. As a boy he explored the lands near his grandfather Walter's ranch, learning about rocks, minerals, mining, and life. One of the things he learned was how to set off a dynamite charge without blowing himself to bits. It was one of the many life-saving skills he acquired as a youth in the rugged outdoors and in his eight years of military training; they would come in handy many years later when he was in the field prospecting and mining.

Fast forward to late 1979. Dave had recently returned from an assignment as a contractor at Kwajalein Island and was working as an executive recruiter ("headhunter") in Reno. An admirer of Israel and the Israel Defense Force, he told me later that he had been mulling over the possibility of joining the Israeli army around that time. However, a confluence of events led to his becoming active in the mining field that winter, and it became his chosen career for the remaining years of his life. A more personal event during this time was meeting me, his future wife, through his friendship with my sister Sheri and her then-husband, Milt. Dave and I were married in Genoa, Nevada, eight months after we were introduced.

Another key element in the circumstances that led to Dave's starting a career in the mining field was the fact that the price of gold had increased dramatically by 1980. This set the stage for Dave to get involved in mining exploration and development full-time. He formed a mining partnership with Milt and his friend Johnny, and the trio staked a claim on a parcel in Gardnerville (about 15 miles south of Carson City) that they named the Poker Flat Mine. The mining prospect didn't pan out, and in late spring 1981 we moved, with our newborn son, to Downieville, a quaint mountain town on the Yuba River in California's Mother Lode country.

Within a few months, Dave began writing a column in the historic local newspaper, the *Mountain Messenger*. Titled *The Miner's Pick*, his weekly column led to his writing articles for the *California Mining Journal* and laid the foundation for his career as a writer and consultant in the mining field. Dave's first love was always mining exploration and prospecting, but writing provided an income base that allowed him to begin acquiring mining prospects (claims) of his own and in partnership with others.

By late spring 1982 we had returned to Nevada; Dave was now writing regularly for the *CMJ* and was a contributor to several other mining publications. By the spring of 1984 he was working in the mining field full-time, prospecting for minerals and eventually consulting while continuing to write, mostly for the *CMJ*.

In 1985, as a member of the Nevada Miners and Prospectors Association (NMPA), Dave began another phase of his mining career: representing the interests of Nevada's small miners and prospectors at the state legislature. It was his love of the freedom and independence afforded by the mining life that drove Dave, now in his mid-forties, to add lobbying to his already full plate: that, and a lifelong inclination to fight injustice. His lobbying efforts at the legislature in behalf of his fellow miners were performed gratis or for minimal compensation. Dave's mission as an advocate for mining and miners was simple but not easy: to help preserve the opportunities that small miners and prospectors in the American West had enjoyed in their chosen field since before the California Gold Rush in the mid-1800s, and to help safeguard their individual and property rights.

Over the next eight years, Dave divided his time between his writing and consulting career, mining and prospecting endeavors, lobbying efforts at the Nevada Legislature, and his family (which, by late 1983, included two sons). But the tremendous stress he was under was taking its toll on him. On a beautiful late-summer Thursday in mid-September 1993, Dave, who had turned 55 the week before, was in his home office finishing some work that he needed to get done before he could head out to his mining property in the Pine Nut Mountains on the upcoming weekend. Suddenly he collapsed, victim of a massive heart attack that claimed his life within minutes despite paramedics' efforts to revive him.

The condolences and tributes to this remarkable man poured in. The late Barbara Vucanovich, representing Nevada in the U.S. Congress at the time, praised Dave's efforts on behalf of the small miner in her remarks for the Congressional Record. On a gray Sunday afternoon at Mormon Station in Genoa, where Dave and I had been married 13 years earlier, family and friends gathered to say their farewells and celebrate Dave's life. Mining associates and others whose individual and property rights he had championed mourned his passing as well. One of them, who was a friend and fellow member of the NMPA, was Sue DeChambeau of Yerington, NV. She paid tribute to Dave and his heroic efforts with these words in her newspaper column, *Mining Is Prosperity* (condensed and edited slightly for reprint here):

> *David vs. Goliath. The 1993 version concerns Dave Parkhurst, champion of the small miner. We lost our champion when Dave died suddenly on September 16. His legacy to us is a package of courage, determination and the spirit to never give up in the struggle to protect our private property rights and the freedom guaranteed us by the U.S. Constitution. Over the many years that he wrote so eloquently about mining in the California Mining Journal, Dave fought the battle to justify the existence of mining, opposed as it is by the so-called environmentalists. He, too, loved being out in the hills prospecting but gave most of his time and energy, aside from time with his family, to writing and lobbying for the cause at the Nevada Legislature.*

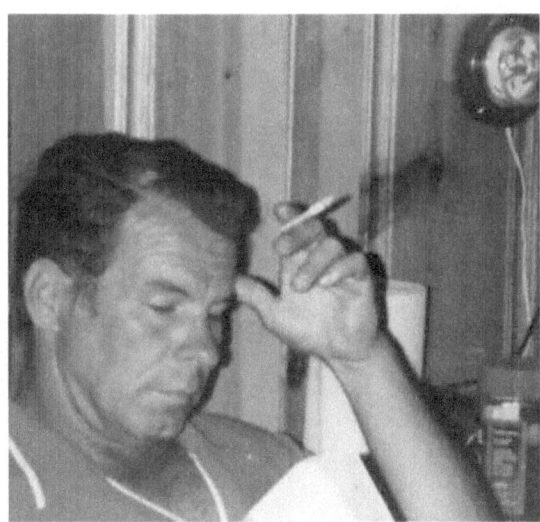

Dave reading at home in Downieville, 1981. He read prodigiously, whether for his work as a mining writer, consultant and advocate or for other purposes. Reading novels, in particular those from the thriller, western, and science fiction genres, was a source of relaxation and escape from the pressures and demands of work in a field that was under constant attack.

APPENDIX A

GOLD PLACERS AND MINERAL DEPOSITS: Their Formation, Deposition and Characteristics is the first of the four volumes that make up the Dave W. Parkhurst Mining Writing Collection. It encompasses four topics:
- How, where and why mineral deposits form.
- The formation and characteristics of placer deposits.
- The nature, characteristics and concentration of gold and where to look for it.
- Prospecting for valuable metals and minerals other than gold.

THE BASICS OF GOING FOR THE GOLD: From Prospecting and Exploration to Small-scale Mining Project, the second volume of the collection, consists of articles on the actual mechanics and processes of prospecting and mining, primarily for gold. It is divided into six main topics:
- Searching for gold.
- Preparing to explore and prospect.
- Prospecting, sampling, and evaluation.
- Recovering the gold and other values.
- Prospecting and mining miscellany.
- Case studies and small mining projects.

A CRITICAL INDUSTRY UNDER ATTACK: The Struggle to Preserve Metals and Minerals Mining Viability in the U.S. is Volume 3 in the set. The three main parts of this book concern these topics:
- The issues, challenges and threats affecting U.S. mining in the 1980s and early 1990s.
- Miners and mining vs. anti-mining extremism.
- Defense of the 1872 Mining Law as amended.

FIGHTING THE GOOD FIGHT: Mining's Battle for Survival in the American West is the last of the four volumes in the Dave W. Parkhurst Mining Writing Collection. It covers five main topics:
- The U.S. government vs. mining.
- Miners and the U.S. Forest Service.
- Implementation of the mining claim holding fee.

- Nevada miners vs. the politicians and anti-mining factions.
- Metals and minerals mining in the Silver State.

The four volumes of the Dave W. Parkhurst Mining Writing Collection are available at Amazon.com and elsewhere. To see more about the books in this collection or about Dave and his, work, please visit **www.PineNutPress.com**.

YEAR OF PUBLICATION IN THE *CMJ*

Article No.	Article Title in Alphabetical Order	Pub. Year
3	Alluvial Fan Gravels	1986
7	Alluvial Placers	1991
8	Beach and Marine Placers	1987
21	Beryllium: The Space Age Metal	1985
12	Buried Placer Deposits	1983
22	Chromium and Chromite	1990
23	Cobalt: A Strategic Metal	1986
9	Desert (Dry) Placers and Gold Deposition	1993
1	Formation of Mineral Deposits, The	1993
4	Formation of New Placer Gold Deposits, The	1992
13	Geology and Technology of Gold, The	1982
10	Glacial, Eolian and Ancient Placer Deposits	1987
14	Gold: Its Character and Concentration	1987
24	Mercury and Cinnabar	1991
2	Metallic Differentiation in Magmas	1987
19	Mineralogy and Geology of Platinum Deposits, The	1989
5	New Gold Placers in Stream Placers After Heavy Runoff	1993
6	Placer Characteristics in the Great Basin and California	1984
17	Placer Gold Mining	1988
20	Prospecting for Diamonds	1991
31	Rare Earth Metals and Minerals, The	1988
11	Residual and Eluvial Placers	1987
16	Separating and Identifying Gold Particles	1991
18	Silver Minerals, The	1991
25	Tantalum and Columbium	1991
26	Tellurium: Properties, Ores and Assay Procedures	1990

Article No.	Article Title in Alphabetical Order	Pub. Year
27	Thorium: Its Geology and Technology	1984
28	Titanium Minerals	1990
29	Tungsten Deposits	1990
15	Types of Gold Ore Deposits	1991
30	Zirconium and Hafnium	1985

APPENDIX C

PUBLICATIONS REFERENCED

Editor's Note: The author listed certain publications as "selected references" in some of the original *CMJ* articles. They are provided below (instead of with the associated article in the text of this book), according to article title and number. Some of the government publications listed may no longer be available from the agencies that originally produced them; for example, the U.S. Bureau of Mines (USBM) closed effective January 1996. For information on obtaining USBM publications, visit the U.S. Geological Survey website or search the Internet for the specific publication. The author may or may not have cited from these sources; if he did, the sources are listed separately, under References.

Alluvial Placers (article 7)
Mineral Deposits, Waldemar Lindgren (1933)
Placer Mining in Nevada, William O. Vanderburg (1936)
Economic Mineral Deposits, Alan M. Bateman (1942)
Placer Examination, Principles & Practice, BLM Technical Bulletin 4

The Mineralogy and Geology of Platinum Deposits (article 19)
Mineral Facts and Problems, USBM 675 (1985)
Manual of Mineralogy & Lithology," James D. Dana (1884)
Mineral Deposits, Waldemar Lindgren (1933)

Chromium and Chromite (article 22)
Mineral Facts & Problems, USBM 675 (1985)
Minerals Yearbook, "Chromium" chapter, USBM
Mineral Deposits, Waldemar Lindgren (1933) (1988)

Cobalt: A Strategic Metal (article 23)
Mineral Facts and Problems, USBM Bulletin 675 (1985)
Mineral Commodity Summaries, USBM (1986)

Mercury and Cinnabar (article 24)
Mineral Deposits, Waldemar Lindgren (1933)
Mineral Facts and Problems, "Mercury" chapter, USBM Bulletin 675 (1985)
Minerals Yearbook, 1988, USBM, "Mercury" chapter

Prospecting for Tantalum and Columbium (article 25)
Mineral Facts and Problems, "Tantalum" and "Columbium" chapters, USBM Bulletin 675 (1985)

"Columbium Availability-Market Economy Countries, " IC 9085, USBM (1986)

Minerals Yearbook, 1989 ed., USBM, "Columbium (Niobium) and Tantalum" chapter

Tellurium: Properties, Ores, and Assay Procedures (article 26)

A Textbook of Fire Assaying, E.E. Bugbee, Wiley and Sons, 1940

Mineral Facts and Problems, "Tellurium" chapter, USBM Bulletin 675 (1985)

Tellurium, 1971 ed., edited by W.C. Cooper, Van Nostrand Reinhold, New York, 437 pp.

Minerals Yearbook, 1987 ed., USBM

Prospecting for Titanium Minerals (article 28)

Mineral Facts and Problems, "Titanium" chapter, USBM Bulletin 675 (1985)

Minerals Yearbook, 1988 ed., "Titanium" chapter, USBM

Mineral Commodity Summaries 1990, USBM

Zirconium and Hafnium (article 30)

"Zirconium and Hafnium in 1984," *Mineral Industrial Surveys,* USBM, Dec. 31, 1984

"Recovery of Byproduct Heavy Minerals from Sand and Gravel Operations in Oregon and Washington," USBM RI 8563, 1981

Mineral Facts and Problems, "Zirconium and Hafnium" chapter, USBM Bulletin 675 (1985)

References

Bateman, Alan M. 1942. *Economic Mineral Deposits.* New York: John Wiley and Sons, Inc.

Boyle, R.W. 1979. "Gold Deposits and Their Geological Classification." Geological Survey of Canada Bulletin 280.

Bugbee, Edward E. 1941. *A Textbook of Fire Assaying.* 3rd ed. New York: John Wiley and Sons, Inc.

Dana, James D. 1884. *Manual of Mineralogy and Lithology.* 3rd ed. New York: John Wiley and Sons, Inc.

Hillebrand, W.F. and Allen, E.T. 1905. "Comparison of a Wet and Crucible-fire Methods for the Assay of Gold Telluride Ores. USGS Bulletin 253.

Lindgren, Waldemar. 1933. *Mineral Deposits.* 4th ed. New York: McGraw-Hill Book Co., Inc.

Lodge, Richard W. 1910. "Notes on Assaying and Metallurgical Laboratory Experiments." 3rd ed. New York: John Wiley and Sons, Inc.

Vanderburg, William O. 1936. *Placer Mining in Nevada.* NBMG Bulletin 27.

Index

alluvial fan gravels
  characteristics of, general, 17-18
  deposits, unique nature of, 17
  economic potential for, 16-17, 18
  geological definition of, 15
  methods for mining of, 18
  minerals deposition and concentration in, 15-16
  precious metals source, 18
alluvial placer deposits, formation of
  deposition of gold in placer concentrations, 41-43
  description of process, 39
  economic placer deposits, 40-43
  mechanical concentration of minerals, 39-40
  pay streaks, 43
  poor man's mine, 41
  potential for gold discoveries, 43
altaite, 137
American Canyon, Nevada, 35
ancient gold placers
  buried deposits in, 58
  characteristics of, 58
  distribution and occurrence of, 57-58
  formation and deposition of, 57
  notable discoveries of, 58
anorthite, 116
antimony, 8, 72, 84, 95, 97, 100, 101, 125, 137, 138, 151
apatite, 108, 132, 160
argentite, 8, 95, 96
arsenic, 72, 84, 95, 97, 98, 100, 101, 125
arsenopyrite, 77, 80, 100
augite, 103

bajada placers, 49
Ballarat, Australia, 57
barite, 61, 124, 160
base metals, 75, 95, 99
bastnasite, 159, 160, 161, 162
batea, 90
bauxite, 61
beach and marine placers
  characteristics of, distinguishing, 47
  discoveries of, past significant, 48
  economic feasibility of mining, 48
  economical deposits, concentration into, 48
  favorable locations of, most, 47
  formation and deposition of, 45
  heavy minerals, mechanical concentration of, 45-46
  less favorable locations of, 47
  mineral commodities produced from, 48
  placer minerals, accumulation of, 46
bertrandite, 113-114
bertrandite deposits, 114
beryl, 8, 113-114
beryllium
  discovery of, 113
  known domestic resources, locations of, 114
  industrial uses of, 113
bismuth, 100, 101, 102, 135, 151
Blake Plateau, 121
bornite, 102, 120
bromyrite, 95, 96
bronzite, 103, 108, 116
bucket-line dredge/dredging, 48, 90, 93, 155
buried placer deposits
  economic deposits, most likely areas of, 67
  gold-bearing deposits, formation and deposition of, 65-66
  hidden deposits, prospecting for, 66-67

buried placer deposits *(continued)*
  locating areas of highest mineral concentration, 66
  methods of prospecting for placer minerals, 66-67
  prospecting for with newer technology, 66
  prospecting for with standard methods, 67
Bushveld Igneous Complex, 103, 117

calaverite, 137
California Gold Rush, 34, 168
California placer deposits, 33-34
Camflo mine, Canada, 81
Canadian gold mines and ore bodies, 80
Carlin ore deposit, Nevada, 81, 126
cassiterite, 18, 33, 39, 100, 142, 152
Central Pacific Basin, minerals in, 121
Central Patricia Mine, Canada, 80
cerargyrite, 95, 96
chalcopyrite, 100, 102, 103, 104, 120
Chinese miners, 20
Choco District, Colombia, 105
chromite
  domestic demand for, 117
  formation, deposition and occurrence, 116-117
  miscellaneous mentions of, 18, 33, 39, 85, 100, 102, 103, 105, 108, 146
  properties and characteristics of, 115
chromium, 11, 99, 115, 116, 117. See also chromite
cinnabar. *See under* mercury
cobalt
  applications and uses of, 119
  concentration and occurrence in nature, 119
  known resources, 120-121
  miscellaneous mentions of, 73, 95, 99, 101, 103, 135
  mining for and concentrations of, 119-120
  Pacific Ocean, mining in, 121-123
  properties and qualities of, 119
  strategic concerns about supply, 119, 120-121
columbite, 18, 33, 39, 129, 130, 133
columbium. *See under* tantalum and columbium
copper: association with other minerals
  beryllium, 113, 114
  gold, 72, 73, 75, 79, 83, 84
  platinum group metals, 99, 100, 101, 102, 105
  silver, 95, 97, 98
Copper Canyon ore body, Nevada, 79
Cordilleran volcanic events, 76, 81
Cortez ore deposit, Nevada, 70
corundum, 100, 109
covellite, 102
crevicing and sniping
  best places to look for precious metals, 25-26
  gold nugget entrapment, 26-27
  harvesting gold using, 22, 25
Crevier deposit, Canada, 133
critical metals and minerals, 11, 99, 115, 119
crystallization, 11, 13, 14, 15, 16, 130, 154

desert (dry) placers
  deposition of dry vs. wet placers, differences in, 49, 50
  fine gold. *See under* fine gold
  float gold. *See under* float gold
  gold concentrations, finding, 49
  gold particles, distribution of, 49, 50
  miscellaneous mentions of, 34, 36-37
  prospecting techniques and sampling, 51
diallage, 103, 104
diamonds
  accessory minerals, 109
  characteristics. properties and composition of, 108
  deposition and occurrence in nature, 109
  description of, 107-108

kimberlite and diamond pipes, 108-109, 110
  occurrence in U.S., 107
  prospecting for, 109
  satellites of, 109
differentiation, metallic. *See* magmas: magmatic differentiation
diopside, 108, 109, 152
dip-box, 90, 91
dry-washing assemblies, 91
dunite, 102

early miners and mining, 19, 20, 23. *See also* old-time miners and mining
electrum, 83, 95, 97
elements
  associated with gold, 72, 75, 84
  in formation of mineral deposits, 7
  tracer, 72, 101
  metallic, 8
  radioactive, 161
  siderophile, 99
eluvial placers. *See* residual and eluvial placers
Emperor telluride leaching process, 136
eolian placer deposits, formation and deposition of, 56-57
epidote, 102, 109
epigenetic view, 77
exploration, mining. *See* prospecting and exploration

Fairbanks, Alaska, ancient gold placer discoveries in, 58
feldspar, 8, 148
fine gold, 20, 48, 51, 71, 86, 91
float gold, 21, 28, 51, 58
fool's gold, 28, 85
forsterite, 100
Fortitude Mine, Nevada, 79
freibergite, 95, 97

garnet, 108, 109, 146, 151, 152, 153
gemstones, 18, 33, 39

Getchell ore deposit, Nevada, 81
glacial placer deposits, 55
gold
  associated with other minerals, 72
  concentration: economic via scavenging, 77; elements associated with, 72, 84
  economic deposits and concentration, 75
  formation and deposition, 76-77
  general characteristics, 71, 75, 83
  industrial uses, 71
  lode deposits, 27, 28, 34, 35, 45, 58, 62, 63, 65, 67, 72
  mining and production, 73
  occurrence in nature, 72, 75, 84
  properties, 72, 75, 83
  purity and fineness, measuring, 71, 76, 83-84
  qualities and value of, 71-72, 76
  with other precious metals, 72, 84
Gold Acres ore deposit, Nevada, 81
Gold Canyon, Nevada, 35
Gold Hill, Utah, 114
gold nuggets
  best places to find larger, 22, 26
  earliest locations found in, 89
  entrapment of, 26
  new crop of, 22, 26-27
gold ore deposits
  Canadian ore deposits and mines, 80, 81
  gold occurrences, eight types of, 79
  known occurrences of, examining as deposit models, 79
  Nevada ore deposits and mines, 79, 81
  Telfer Mine in Australia, 80
  Witwatersrand Basin in South Africa, 81
gold placers. *See* individual placer types
gold processing and recovery, 19, 73, 89-93, 110
gold rockers, 91
gold, metallic, 8

gold, separating and identifying
  appearance, 83, 85, 86
  concentration with other valuable minerals and metals, 85
  disseminated particles, 84
  distinguishing characteristics, 85-86
  fool's gold, distinguishing from gold, 84-85
  identification tools and methods, 85-86
  magnets, using for separation, 86
  miner's gold pan and panning, 72, 84, 89, 90, 126
  particles, separating from host minerals, 84
  prospecting by novices and beginners, 83, 84
Goodnews Bay, Alaska
  beach placer gold discovery, 48
  platinum placer discovery, 105
gorceizite, 109
graphite, 75
gravel processing plant, 90, 93
gravity concentration method, 39, 73, 86, 89, 93, 149, 153, 156
Great Basin and California placers
  California gold placers, 33-34
  desert (dry) placers, 34, 35-37
  definition of placer, 33
  Nevada gold placers, 35
  placer deposits, distribution and characteristics of, 33
  placer mining, business of, 37-38
Great Basin range, 35
Great Dyke, Zimbabwe, 117
ground sluicing activity, 92

hafnium. *See* zirconium and hafnium
hardrock deposits, 27, 62, 83, 132, 149
heavy/high water runoff, 21, 22, 25-29, 42, 49, 52
hematite, 85
hessite, 95, 137
hidden lode deposits, 63
hillside placers, 62
horn silver, 96

hornblende, 103
hydraulicking, 92

identifying gold, 84-87
igneous syngenetic deposits, 11
ilmenite, 18, 33, 39, 48, 85, 100, 108, 109, 116, 142, 145-149, 155
industrial diamonds, 107
iridium, 99, 105
iron ore, 61, 120
iron pyrite, 84, 85. *See also* fool's gold
ixiolite, 129

Jos Plateau, Nigeria, 130

kalgoorlite, 137
Kapuskasing-Moosonee High, 131
karat, 71
Kola Peninsula, Lovozero alkali massif, 132
krennerite, 137
kyanite, 61, 156

lacustrine deposits, 52
Lamaque Mine, Canada, 81
lollingite, 100
long tom, 90, 91
loparite, 129, 132, 152, 160
Lovozero alkali massif, Kola Peninsula, 132
Lydenburg District, 104

magmas
  composition of, 11-12
  differentiation of intrusive igneous masses, 11
  economic deposits, original source of, 12
  formation and movement of, 12
  magmatic differentiation, 14
  mineral crystallization, 11, 13
  parent magmas, 11, 12, 13, 14
  strategic and critical minerals, source of, 11
magmatic ore bodies
  description of and other terms for, 11

Index

economic mineral deposits, 12, 13, 14
magnetite, 39, 85, 100, 103, 109, 116, 132, 146, 147, 149
manganese, 61, 121, 122, 136, 152, 153
Mercur Mine, Utah, 127
mercury
  byproduct of gold and silver mining in Nevada, 126
  cinnabar, 8, 85, 123, 124, 125, 126
  formation and deposition of, 123-124
  indicator of precious metals, 126
  properties and characteristics of, 123
  prospecting for, 126
Merensky horizon-type reef forms, 103
metallic differentiation in magmas. *See* magmas, magmatic differentiation
mica, 8, 28, 85
microlite-pyrochlore mineral series, 130
miner's gold pan, 72, 84, 89, 90, 126
mineral deposits formation
  deposition, conditions affecting, 7
  economic mineral deposits, 9-10
  elements, role in, 7
  precious metal vein deposits, 8
  transportation and deposition, processes involved in, 7-9
  veins and pockets, 7
mineral deposits, identifying economic, 10
miners, Chinese, 20
mining equipment, 19, 37
mining, placer. *See* placer mining
Mojave desert placers, 34
monazite, 18, 33, 39, 48, 109, 141, 142, 146, 147, 149, 155, 159, 160, 161, 162
molybdenite, 100
moraines, 34, 49, 55-56
Mother Lode, 34, 167

nagyagite, 137
Nevada mines and ore bodies, 79, 81, 126-127
nickel, 61, 71, 73, 99, 100, 101, 102, 103, 117, 119, 120, 121, 122, 135, 136
Niobec ore body, Canada, 131, 132
Nome, Alaska, beach gold placer discovery, 48
norite, 103, 104

Oka deposit, Canada, 132
old-time miners and mining, 21, 26, 35. *See also* early miners and mining
Olinghouse, Nevada, 35
olivine, 100, 102, 103, 108, 109, 117
opal, 123, 124
open-pit mining, 18, 73, 162
Oregon beach placer gold, 34, 48, 105
Osceola, Nevada, 35
osmium, 99

Pacific Ocean, mining in, 117, 121-122
palladium, 71, 83, 99, 100, 101, 102, 103, 104
pandaite, 129, 131, 132
pentlandite, 100, 103
peridotite, 100, 102, 103, 105, 107, 108, 116
petzite, 137
Pickle Crow deposit in Canada, 80
placer deposits
  characteristics and definition of, 33
  types of. *See* individual placer types
placer deposits, new
  concentrations of gold particles, 22
  early miners' gold losses, 19-20
  economic potential for, 19
  erosion and water runoff, effects of, 20-21
  gold crop, harvesting annual, 22
  gold values, deposition and distribution, 19-21
  heavier placer minerals, potential for, 22-23
  ideal conditions for deposition of, 21-22
  low-grade deposits, mining larger, 22

placer deposits, new *(continued)*
  minerals, concentration of, 21
  reconcentration of gold values, 19
placer gold, mining of and prospecting for
  early history of, 89
  methods and devices used in, 89-93
  placer minerals recovery, 90-93
  placer mining described, 89-90
placer mining
  business of, 38
  equipment, 37
platinum deposits
  characteristics of, 99-101
  formation and deposition of, 100-102
  notable discoveries and geographic occurrence of, 102-105
  occurrence in nature, 99-100, 101, 102
  platinum group. *See* platinum group metals
  platinum in placer deposits, 104-105
  types of occurrences, 99-100
platinum group metals, 11, 27, 33, 39, 41, 48, 57, 81, 85, 99, 100, 101, 103
plumbojarosite, 102
podiform ore deposits, 116, 117
Point Orford, past platinum placer production, 105
Porcupine deposit, Canada, 80
polybasite, 95, 97
prospecting and exploration
  assaying, 84, 136, 138-140
  exploration methods, sophisticated, 18, 35, 63, 66
  field examination and testing, 86-87
  miner's gold pan, 72, 84, 89, 90, 126
  novice and beginner prospectors, 83, 84, 86
  panning and sampling, 28, 37, 51, 84, 85, 126, 147
  sampling, improper/incorrect, 136, 139
  techniques and methods, 51, 66, 83, 89-93, 110, 126

tools and equipment, 19, 22, 37, 51, 63, 85, 89-93
proustite, 95, 97
pyrargyrite, 95, 97
pyrochlore, 129, 131-132
pyrrhotite, 100, 102, 103
pyroxenite, 102, 105, 116

quartz
  gangue material, 72, 80, 123, 124
  veins, 80, 100, 102, 104, 152
quartzite, 60, 85, 141
quicksilver, 123, 125

Rambler Mine, Wyoming, 102
rare earth metals and minerals
  applications, uses and demand for, 159, 162-163
  domestic consumption and production, 161
  exploring for, 162
  lanthanide series, 131, 159
  mining and recovery processes, 162-163
rare earth elements, subgroup classifications of, 159
rare earth metals: and yttrium, 159; primary mineral sources of, 159
rare earth oxides, 142, 160
rare earth-bearing minerals, most important, 160
reconcentrated placer deposits. *See* new placer deposits
Reichert cone concentrator, 148
Renabie ore body, Canada, 81
REO, 160-161
residual and eluvial deposits
  definition, description of and distinction between the two, 59
  disintegration and decomposition of, 59-60
  economic concentrations, conditions necessary for, 61-62
  economically viable, 61
  effects of weathering, varying, 59

physical character and characteristics of, 62-63
potential for discovery of, 63
tropical regions, minerals produced in, 60
rhodium, 99, 104
Russian gold, 82
Russian platinum deposits, 102
ruthenium, 99, 104
rutile, 18, 33, 39, 85, 109, 116, 130, 142, 145, 146, 147, 148, 149

Salt Chuck Mine, Alaska, 102
scheelite, 18, 33, 85, 151, 152, 153
selenium, 135
sericite, 81
serpentine, 102, 105, 108, 116, 123
Serpentine Dike, California, 116
siderophile elements, 99
Sierra Nevada glacial stream placers, 34
silver
  association with gold or the basic metals, 95
  copper-bearing minerals important as sources of, 97-98
  deposition and distribution of, 98
  important silver ores: characteristics and notable occurrences of, 96-97
  known world reserves, 98
  main silver minerals of economic importance, 95
  occurrence of in U.S., 95
  properties of, 95
silver glance, 96
skarn-type deposits, 79, 151, 152
sluice box, 19, 89, 90, 92, 110
Solowioff Mountain, 102
sniping. *See* crevicing and sniping
South African platinum deposits, 101, 103
sperrylite, 102
spinel group, 116
spinel minerals, 100, 109
Spor Mountain, 114

spring runoff, 21, 25, 26, 36, 42, 92. *See also* water runoff
Spring Valley, Nevada, 35
St. Lawrence River Fault, columbium deposits, 131
staurolite, 109, 146
stephanite, 95, 97
stibnite, 8, 123, 125
Stillwater Complex, Montana, 101, 103, 116, 117, 119, 120
strategic metals and minerals, 11, 99, 115, 119, 120
stratiform ore deposits, 116, 117
stream placer gold deposits
  best places to look for gold, 25-26, 27, 28
  crevicing. *See* crevicing and sniping
  deposition and concentration of placer gold, 26-27
  gold concentration, effects on by changing geological conditions, 29
  known placer gold occurrences in U.S., 27
  likely areas to prospect, 26-27
  new gold placers, effects of high water runoff on formation of, 25
  old gold-producing districts, good places to pan in, 28
  panning and sampling, 28-29
  sniping. *See* crevicing and sniping
strontium, 132, 160
struverite, 130
sylvanite, 137

tantalite, 18, 33, 39, 129, 130
tantalum and columbium
  concentrations, deposition and distribution of, 129-133
  economic mineral concentrations, 129, 130
  geographic occurrence in U.S., 129
  geographic occurrence in the world, 130, 131, 132, 133
  occurrence in nature and together, 129

tapalpite, 137
Telfer Mine, 80
tellurides. *See under* tellurium
tellurium
   assaying problems and solutions, 138-140
tellurium
   common telluride ores, 136-137
   mining and processing of, 136
   native state properties and distribution, 135
   occurrence, deposition and distribution of, 135, 136
   properties of, 135
   tellurides, 135, 136, 137, 138, 139
   testing and analysis of, 138
tennantite, 98
tetrahedrite, 97
Thor Lake deposit, Canada, 133
thorianite, 141
thorite, 141, 142
thorium
   availability of, 143
   finished products, principal, 143
   industrial uses of, 143
   mining and recovery of, 142
   mining operating factors, 143
   occurrence, deposition and sources of, 141-142
   properties of, 141
   safety and environmental aspects related to, 143
   uranium and rare earth elements associated with, 131, 141, 155, 161
thorium compounds, 143
titanium minerals
   commercial value of, 145
   commercially viable ore deposits, 146
   deposit types found in, 147
   economic deposits of, 147-148
   important and most common, 145-146
   minerals closely associated with, 146
   mining and processing methods, 148-149
   occurrence in nature, 145
   occurrence, deposition and distribution, 146-147
   properties and composition of, 145-146
   prospecting, sampling and panning for, 147-148
   radioactive minerals, closely associated with certain, 146
   visible signs of, 147
topaz, 109
tourmaline, 85, 109
Tulameen District, British Columbia, 105
Tulameen River, British Columbia, 102
tungsten deposits
   contact metamorphic deposits, close association with, 151
   formation, occurrence, and primary sources of, 151-153
   economic tungsten ores, the primary, 152
   known world reserves of, 152
   mining and processing of. 153-154
   scheelite characteristics and occurrence of, 152
   tactite deposits, primary source of, 152. *See also* skarn-type deposits
   wolframite characteristics and occurrence of, 152, 153

Ural Mountains in Russia, 105, 117
uranium, 81, 131, 141, 142, 143, 160, 161-162

Vauquelin, 113

water runoff
   effects of on minerals deposition, 16, 17-18, 39, 59, 62
   heavy/high, 21, 22, 25, 26, 27, 29, 42, 49, 52
   spring runoff, 21, 25, 26, 36, 39, 42, 92
Waterberg District of central Transvaal, 104

Western Australia
  economic eolian deposits in, 56
  tellurium in, 135, 137
  zircon in, 157
Witwatersrand Basin, South Africa, 81
Witwatersrand goldfields, 77
Witwatersrand placers, 104
wolframite, 18, 85, 151, 152, 153

xenotime, 159, 160, 161, 162

Yellow Pine District, Nevada, 102
yttrium, 142, 155, 159, 160, 161, 162

zircon, 18, 39, 48, 85, 109, 116, 146, 148, 149, 155-156, 157
zirconium and hafnium
  as byproduct of titanium and tin mining, 157
  domestic production of, 157
  hafnium use in strengthening of alloys, 157
  industrial applications of, 156-157
  mining methods and processes, 155-156
  occurrence in nature, 155
  sources of, primary, 155
  world production of, 157

Editor Contact Information

Susan Lee (Sue) Parkhurst is the publisher/editor of Pine Nut Press in Minden, Nevada. To learn more about the publications and services offered by Pine Nut Press, or to see more of Dave's writing or his project photos, please visit **www.pinenutpress.com**.

The print version of the Dave W. Parkhurst Mining Writing Collection may be purchased from Amazon.com. E-book versions of the volumes are planned and will also be available at Amazon.com.

www.ingramcontent.com/pod-product-compliance
Lightning Source LLC
Chambersburg PA
CBHW020651220526
45464CB00001B/386